Control of Cardiac Output

Integrated Systems Physiology: From Molecule to Function

Editors

D. Neil Granger, *Louisiana State University Health Sciences Center*
Joey P. Granger, *University of Mississippi Medical Center*

Physiology is a scientific discipline devoted to understanding the functions of the body. It addresses function at multiple levels, including molecular, cellular, organ, and system. An appreciation of the processes that occur at each level is necessary to understand function in health and the dysfunction associated with disease. Homeostasis and integration are fundamental principles of physiology that account for the relative constancy of organ processes and bodily function even in the face of substantial environmental changes. This constancy results from integrative, cooperative interactions of chemical and electrical signaling processes within and between cells, organs and systems. This eBook series on the broad field of physiology covers the major organ systems from an integrative perspective that addresses the molecular and cellular processes that contribute to homeostasis. Material on pathophysiology is also included throughout the eBooks. The state-of the-art treatises were produced by leading experts in the field of physiology. Each eBook includes stand-alone information and is intended to be of value to students, scientists, and clinicians in the biomedical sciences. Since physiological concepts are an ever-changing work-in-progress, each contributor will have the opportunity to make periodic updates of the covered material.

Published titles

(for future titles please see the website, www.morganclaypool.com/page/lifesci)

Capillary Fluid Exchange: Regulation, Functions, and Pathology
Joshua Scallan, Virgina H. Huxley, and Ronald J. Korthuis
2010

The Cerebral Circulation
Marilyn J. Cipolla
2009

Control of Cardiac Output
David B. Young
www.morganclaypool.com

ISBN: 9781615040216 paperback

ISBN: 9781615040223 ebook

DOI: 10.4199/C00008ED1V01Y201002ISP006

A Publication in the Morgan & Claypool Life Sciences series

INTEGRATED SYSTEMS PHYSIOLOGY: FROM MOLECULE TO FUNCTION

Lecture #6

Series Editors: D. Neil Granger, Louisiana State University; Joey Granger, University of Mississippi

Series ISSN Pending

Control of Cardiac Output

David B. Young
University of Mississippi Medical Center

INTEGRATED SYSTEMS PHYSIOLOGY: FROM MOLECULE TO FUNCTION #6

ABSTRACT

Although cardiac output is measured as the flow of blood from the left ventricle into the aorta, the system that controls cardiac output includes many other components besides the heart itself. The heart's rate of output cannot exceed the rate of venous return to it, and therefore, the factors governing venous return are primarily responsible for control of output from the heart. Venous return is affected by its pressure gradient and resistance to flow throughout the vascular system. The pressure gradient for venous return is a function of several factors including the blood volume flowing through the system, the unstressed vascular volume of the circulatory system, its capacitance, mean systemic pressure, and right atrial pressure. Resistance to venous return is the sum of total vascular resistance from the aortic valve to the right atrium. The sympathetic nervous system and vasoactive circulating hormones affect short-term resistance, whereas local tissue blood flow autoregulatory mechanisms are the dominant determinants of long-term resistance to venous return. The strength of contraction of the heart responds to changes in atrial pressure driven by changes in venous return, with small changes in atrial pressure eliciting large changes in strength of contraction, as described by the Frank–Starling mechanism. In addition, the autonomic nervous system input to the heart alters myocardial pumping ability in response to cardiovascular challenges. The function of the cardiovascular system is strongly affected by the operation of the renal sodium excretion–body fluid volume–arterial pressure negative feedback system that maintains arterial blood pressure at a controlled value over long periods. The intent of this volume is to integrate the basic knowledge of these cardiovascular system components into an understanding of cardiac output regulation.

KEYWORDS

blood flow autoregulation, blood volume, circulatory shock, heart failure, hemorrhage, mean systemic pressure, myocardial contractility, pressure gradient for venous return, vascular capacitance, vascular resistance, venous return

Preface

We learn early in our education that blood is pumped by the heart through the arteries, first to the lungs and then to the rest of the body, returning to the heart through the veins. However, the circulation that now seems so obvious was not understood until the eighteenth century when Hale first described it. He recognized that the heart was the center of the circuit, providing the motive force propelling blood at high pressure into the arteries. For the next several hundred years, physicians and physiologists inferred from the heart's power and position at the center of the system that it controlled the flow of blood throughout the body, and today, the majority of physicians still may believe cardiac output to be regulated by the pumping strength of the heart. However, during the 1950s, Arthur Guyton and coworkers conducted an extensive series of experiments that, together with what had been learned earlier by Starling, Wiggers, and many others, led to the conclusion that, in fact, the normal heart could do little more than pump what returned to it from the veins; it had no control over the rate of output but could only respond to the flow of blood into its chambers. Guyton concluded that cardiac output was controlled instead by the factors that regulated the flow of blood from the body back to the heart. While this may seem to be a straightforward hypothesis, the circulatory mechanisms involved in mediating the control are complex and interrelated with other systems, and consequently, the full understanding of cardiac output regulation requires command of many aspects of physiology.

The publications of Guyton and coworkers and their contemporaries beginning approximately 50 years ago essentially provided all the data required to strongly support this hypothesis, and their conclusions were formalized in 1973 in the second edition of *Circulatory Physiology: Cardiac Output and Its Regulation* by Guyton et al. [1]. My purpose in this publication is to reacquaint the reader with the concepts they had developed, concentrating less on the techniques and data from the original experiments and more on the resulting concepts. In addition, more recent significant elaborations and modifications of the original work are included, along with results of simulations designed to assist in appreciating the work. I consider this to be worthwhile because a full presentation of the concepts is no longer available, and to this day, they are valid and essential in understanding circulatory physiology and pathology.

The concepts developed over the past 50 years are the basis of a complex mathematical model, *Digital Human*, developed by Thomas G. Coleman, a coauthor of the aforementioned book. The equations that comprise the cardiovascular section of the model were derived directly from the concepts and logic developed earlier by Guyton, Coleman and associates, and therefore, its simulations are faithful representations of the theories of operation of the cardiovascular control system based on that logic. I encourage the reader to take advantage of Professor Coleman's outstanding work and use *Digital Human* to investigate additional aspects of cardiovascular control (go to http://digitalhuman.org).

Physiologists and clinicians learned many things from Arthur Guyton; among the most important may be that worthwhile understanding of physiology requires the study of more than tissues or organs, but instead analysis of complete systems, with time as a variable. The lesson is well illustrated by the concepts he developed concerning cardiac output regulation.

Contents

CHAPTER 1

Introduction

Life began in the primordial ocean, the primitive cells surrounded by a nearly infinite aqueous medium supplying the needs for the single-celled inhabitants. As the complexity of life forms increased, cells could not be in contact with the exterior bathing medium, and consequently, a means to circulate the interior medium between the exterior and the cells on the interior was a necessary feature of even the first complex organisms. In man and other mammals, the means of circulation has become extremely complex, allowing for the development of tissues with large metabolic demands all located within a few microns of capillaries perfused with blood replete with oxygen and the cells' metabolic substrates. The heart must pump blood through many kilometers of vascular channels so that exchange can occur between plasma and extracellular fluid across hundreds of square meters of capillary surface area.

In healthy individuals, cardiac output can increase from 5 to approximately 20 L/min. At rest, man's cardiac output is approximately 5 L/min for a 70-kg person or, stated differently, the cardiac index of a healthy man with a surface area of 1.7 m^2 is approximately 3 L/min/m^2 of body surface. Cardiac output varies in approximate proportion to lean body mass. Normal blood volume is about equal to the amount of blood pumped from the heart each minute, and consequently, extensive exchange and mixing between the extracellular fluid and the plasma can be maintained. In healthy individuals, cardiac output can increase at least 3-fold to between 15 and 20 L/min. Such an increase can occur in response to maximal metabolic demand by large working muscles during extremely vigorous exercise, as would occur if one was running to escape mortal danger or competing in an especially intense sporting event. An average person can improve maximal cardiac output with training, but only by about 10–20%. However, world-class athletes can achieve maximal rates approaching 40 L/min; the apparent incongruity is probably due to a genetically determined greater than normal heart size along with the intensive long-term training undertaken by exceptionally able competitors. A person's cardiac output is maximal in the early 20s and tends to decline later in life. The changes throughout life are probably secondary to changes in the body's metabolic rate, unless the heart is severely weakened by disease.

In healthy individuals, metabolic rate and cardiac output are closely correlated. This was noted by Guyton and associates [1], who compiled previously published oxygen consumption and cardiac output data obtained from subjects working at rates spanning an order of magnitude. Figure 1.1 illustrates the parallel increases in cardiac output and oxygen consumption over the wide range of metabolic demand, both in sedentary individuals and in trained athletes.

The close relationship between the variables is evident from these data, and it can be observed in response to other conditions in which metabolic rate changes. During digestion of a meal, the increased metabolic activity of the liver and gastrointestinal organs is associated with as much as a 25% elevation in cardiac output; during sleep, cardiac output decreases by about 25%, and in response to anxiety, elevation of metabolic rate driven by the sympathetic nervous system measurably raises cardiac output. During cold exposure, the muscle metabolic activity of shivering raises cardiac output, while in response to elevation of body temperature, cardiac output also increases to provide increased blood flow to the skin for thermoregulatory purposes. The interrelationship between cardiac output and oxygen consumption is apparent in both trained athletes and untrained individuals and in subjects before and after aerobic conditioning [2].

In two prominent organs, metabolic demand may not be the only factor associated with demand for blood flow. The two kidneys each receive about 10% of resting cardiac output, more than what is required to meet the metabolic requirements of their tissues. Instead, renal blood flow is maintained at a higher level to supply sufficient flow to the glomeruli to filter and excrete the metabolic waste products of the whole body. Skin blood flow also varies in a manner that is largely

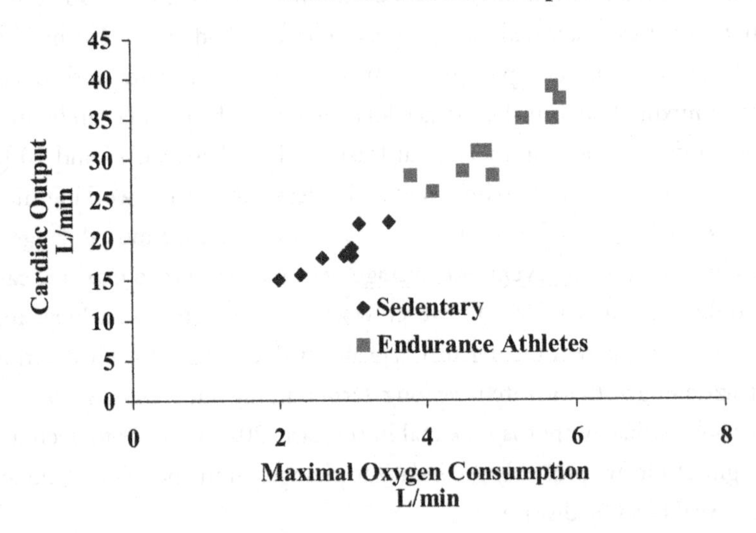

FIGURE 1.1: Correlation between maximum oxygen consumption and maximum cardiac output in trained athletes and untrained subjects. From reference [2].

independent of its metabolic needs. The overriding determinant of flow to the skin is the body temperature. Blood flow to the body surface allows heat loss from the body, which may be desirable or undesirable, depending on the circumstance. When core body temperature is well below normal, or more precisely below the set point of the negative feedback control system that regulates body temperature, skin blood flow may be reduced to a few hundred millilitres per minute and in localized regions to near zero for intermittent periods. Conversely, when core temperature is several degrees above the normal level, skin blood flow can rise as high as 6 L/min.

1.1 FUNCTIONAL CHARACTERISTICS OF THE VASCULAR SYSTEM

The anatomy and basic function of the vascular system are probably well known by anyone interested in this topic. However, briefly reviewing some of the more subtle aspects of the functional characteristics will be helpful in appreciating the remainder of the presentation.

Ohm's law can be applied to the study of cardiac output regulation. Although Ohm's law was originally formulated to describe the flow of electric current, it can also be usefully adapted to the study of blood flow:

$$\text{Flow} = \text{change in pressure/resistance}$$

where "flow" is the flow of blood through a system of tubes, "change in pressure" is the difference in pressure measured at the beginning and end of the system, and "resistance" is the impediment to flow encountered by the blood moving through the system. For the systemic circulation, the flow is equal to the pressure in root of the aorta minus the right atrial pressure divided by the systemic resistance. Pulmonary flow is calculated from the pressure difference between the pulmonary artery and the left atrium divided by pulmonary resistance. The resistances of the two circulations are the sum of all resistances throughout the systems, values that cannot be measured but can be calculated if systemic flow (cardiac output) and the differences in pressure are known.

Pouseuille's law is an extension of Ohm's law that is more useful in studying the cardiovascular system:

$$\text{Flow} = [(\pi)(\text{change in pressure})(\text{radius}^4)]/[8(\text{viscosity})(\text{length})]$$

where π is the mathematical constant (3.14159), "change in pressure" is the difference in pressure at the beginning and end of a tube or vessel, "radius" is that of the tube, "viscosity" is that of the fluid or blood, and "length" is that of the tube. Significantly, flow is proportional to the fourth power of the radius; therefore, doubling the radius of a vessel will cause a 16-fold increase in flow.

Mean blood pressure in the systemic circulation normally ranges from 100 mm Hg in the root of the aorta to approximately 0 mm Hg in the right atrium. Arterial pressure is much higher in the systemic circulation than in the pulmonary circuit, averaging approximately 100 mm Hg versus approximately 20 mm Hg in the pulmonary circulation. The majority of the reduction in pressure in the systemic side occurs in the arterioles due to the high resistance of these small vessels. Table 1.1 lists the total cross-sectional areas and pressures in the vessel types of the systemic circulation.

The total blood volume is approximately 7% of the body weight and is distributed throughout the segments of the circulatory system as indicated in Table 1.2. Noteworthy is the large percentage of blood present on the venous side of the system.

The capacity of a segment of the circulation to contain blood at a given pressure level is termed *capacitance* or *compliance*:

Vascular capacitance = change in volume/change in pressure

The thin-walled veins are more distensible than the thicker walled arteries, and the veins contain more blood. Consequently, the veins' capacitance is much greater (18 times greater) than that of the arteries [3].

Mean systemic pressure refers to the pressure measured anywhere in the systemic vascular system immediately after the cessation of circulation when all pressures have equalized. Actual measurement of this pressure is difficult because several seconds are required for the pressures through the system to equalize after the heart has stopped, and within approximately 7 s, activation of sympathetic nervous system reflexes begins to constrict the blood vessels. However, the pressure can be determined from experiments designed to minimize artifacts, which Guyton and coworkers performed in an extensive series of studies beginning in 1954 [4]. Interestingly, their initial publication of this work included venous and arterial pressure measurements recorded on a kymograph. The mean value that they obtained in anesthetized dogs was 7.0 mm Hg. Mean circulatory pressure is the pressure measured anywhere in the vascular system, including the pulmonary circulation, when the heart has stopped and arterial and venous pressures have equalized. They found that, in the anesthetized dog, the pressure averages 6.9 mm Hg. Mean pulmonary pressure can be determined similarly, and it averages a few millimeters of mercury less than mean systemic pressure.

1.2 LOCAL TISSUE AUTOREGULATION OF BLOOD FLOW AND ITS IMPORTANCE IN DETERMINING CARDIAC OUTPUT

Cardiac output is distributed throughout the organs and tissues of the body, generally according to metabolic demand. In most conditions, the metabolic need for oxygen is the dominant factor

TABLE 1.1: Cross-sectional areas and pressures of blood vessels.

	AREA (cm^2)	PRESSURE (mm Hg)
Aorta	2.5	100
Small arteries	20	95
Arterioles	40	85–35
Capillaries	2500	35–10
Venules	250	10–5
Small veins	80	5–2
Venae cavae	8	2–0

affecting a tissue's blood flow resistance. Each gram of tissue has the capability to control the flow of blood through its microcirculation by altering the resistance of the small arteries and arterioles supplying its capillary network. The physiology of the microcirculation is an important and fascinating discipline, but it lies beyond the scope of this presentation. Another volume of this series fully presents the topic. For our purposes, it is sufficient to state that, while a variety of factors are capable of affecting microcirculatory resistance, the oxygen concentration of the local extracellular fluid has the greatest long-term effect.

Tissue blood flow is regulated by a local feedback control system. When the metabolic activity of a region of tissue increases, its utilization of oxygen increases, thus increasing the rate of removal of oxygen from the local extracellular fluid below the rate of delivery from the blood in the capillaries. Consequently, oxygen concentration in the extracellular fluid declines. Local mechanisms

TABLE 1.2: Distribution of blood volume throughout the vascular system (%)

Heart	7
Pulmonary circulation	9
Arteries	13
Arterioles and capillaries	7
Veins, venules, and venous sinuses	64

sensitive to extracellular fluid oxygen concentration respond with actions that cause vasodilation of the arterioles supplying the capillaries in that region of the tissue. With the subsequent reduction in resistance, blood flow to the capillaries increases and oxygen delivery rises, allowing the rate of oxygen diffusion into the extracellular fluid to increase. The process continues until the rate of entry of oxygen returns the extracellular fluid oxygen concentration close to the desired level or set point concentration. This negative feedback control system regulates tissue blood flow throughout the body, acting within seconds to maintain oxygen concentration near the set point level in response to changing metabolic demand. The control system is referred to as the short-term tissue autoregulatory mechanism. A related long-term system also operates over periods of days and weeks; in response to prolonged reductions in tissue oxygen concentration, growth of additional capillaries and other microcirculatory vessels takes place, stimulated by mechanisms of the long-term autoregulatory system. The increased capillary density provides a long-term increase in oxygen delivery to the tissue.

1.3 SUMMARY

Cardiac output is equal to the summated blood flow throughout all tissues of the body. Therefore, the factors that control local tissue blood flow strongly affect the regulation of cardiac output. Consequently, the requirement of the body's tissues for oxygen is a prominent determinant of cardiac output. The mechanisms linking the concepts in these three sentences comprise most of what is contained in the remainder of this presentation.

• • • •

CHAPTER 2

Venous Return

Venous return refers to the flow of blood from the periphery back to the right atrium, and except for periods of a few seconds, it is equal to cardiac output. Because clinicians and investigators have long observed that factors affecting primarily the venous side of the circulation can have profound influence on cardiac output, mechanisms governing the flow of blood to the heart have been studied in some depth. However, full understanding of the venous side has been challenging because of the complex nature of some of its characteristics.

Guyton recognized the importance of determining the role of both mean systemic pressure and right atrial pressure in controlling venous return, and measuring both accurately proved to be very difficult. Mean systemic pressure can only be measured when pressures throughout the systemic circulation have come to equilibrium after the heart has stopped beating. He and his co-workers developed techniques to temporarily stop the heart by electrical fibrillation or other means while blood was mechanically pumped rapidly through an arterial to venous shunt in order to bring the arterial and venous pressures to equilibrium in less than 7 s [4]. They found that the technique could be repeated many times without affecting the value of mean systemic pressure. Analysis of the effects of changes in the right atrial pressure on venous return required a means to vary the right atrial pressure in a controlled manner. Because right atrial pressure is a function of output from the right atrium into the right ventricle and flow of blood into the atrium from the vena cava, they had to develop a preparation in which outflow could be augmented by mechanically pumping from the atrium through an extracorporal shunt into the pulmonary artery or through a heart lung bypass machine to the aorta. By varying the rate of flow around the atrium, they were able to reduce right atrial pressure in a controlled manner while measuring cardiac output in the systemic circulation, from which they knew the value of venous return. They could also augment flow into the atrium from a reservoir, thereby increasing right atrial pressure. Some of the experiments were conducted with closed chest animals that were breathing again negative pressure as low as −10 mm Hg, lowering the intrathoracic and right atrial pressures to the lowest physiological levels. These extensive series of experiments laid the groundwork for an in-depth understanding of the basic factors controlling venous return.

2.1 DETERMINANTS OF VENOUS RETURN

Everywhere in the body, pressure gradients and resistances determine blood flow rate. When considering venous return, the pressure gradient is mean systemic pressure minus the right atrial pressure, and resistance is the total peripheral vascular resistance. It is probable that some of the difficulties associated with conceptualizing and measuring mean systemic pressure have contributed to the challenges in understanding venous return.

Mean systemic pressure is affected by blood volume and vascular tone. When blood volume is normal, mean systemic pressure is approximately 7 mm Hg [3, 4]. If mean systemic pressure is measured after blood volume has been changed rapidly in steps above and below normal while sympathetic nervous system activity has been blocked, a volume–pressure relationship is obtained. An example of such a curve is illustrated by the solid line in Figure 2.1. Several of its characteristics are significant: first, the relationship is approximately linear within physiological limits, and second, it is highly sensitive to changes in volume, with a 15% reduction in blood volume (approximately 1 L in man) decreasing mean systemic pressure from 7 to 0 mm Hg. The blood volume at which mean systemic pressure is 0 mm Hg is termed the *unstressed vascular volume* of the system. Reducing blood

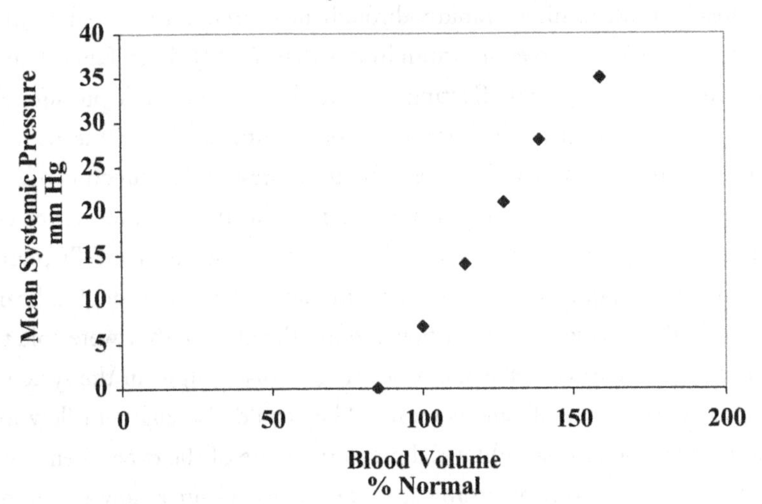

FIGURE 2.1: Relationship between percentage changes in blood volume and mean systemic pressure measured with reflex responses blocked. Unstressed vascular volume is the blood volume at which mean systemic pressure is 0, approximately 85% of normal blood volume. Data are from references [3] and [4].

volume below the unstressed vascular volume does not result in further reduction in mean systemic pressure.

The sympathetic nervous system and locally acting and circulating vasoactive hormones affect vascular smooth muscle tone. Increasing vascular tone shifts the volume–pressure curve to the left without significantly affecting the slope. When vascular tone increases, unstressed volume decreases and mean systemic pressure increases for each level of blood volume. Conversely, totally blocking the sympathetic nervous system or otherwise reducing vasomotor tone has been shown to shift the curve to the right in a parallel manner.

The vessels on the arterial side have much less capacity and are much less distensible than the veins, and consequently, the characteristics of their volume–pressure relationship differ markedly from those of the venous side; the unstressed arterial vascular volume is approximately 0.4 versus 4.0 L for the venous side, and the arterial slope is 33 versus 7 mm Hg/L. In addition, the arterial vessel wall is more responsive to sympathetic nervous system innervation and vasoactive hormones.

Several sites in the vascular system have large reservoir capacities. Portions of the vascular system have a large capacitance, that is, they can gain or lose large volumes of blood with little change in pressure. Therefore, as pressure within other portions of the venous system increases or decreases, large volumes of blood can move into or out of these reservoirs, buffering changes in pressure throughout the vascular system. Smooth muscle of the vascular walls of some of the vessels in these sites can contract in response to sympathetic stimulation and circulating vasoconstrictor substances, significantly decreasing their capacitance and causing additional blood to be translocated to other portions of the circulation. Large veins in the abdomen and thorax are especially effective reservoirs, as are the sinuses of the spleen and liver. The vascular plexuses of the skin can also function as reservoirs. Blood flow into the skin is highly responsive to catecholamines released from the sympathetic nerves innervating the resistance vessels of the skin, the constriction of which decreases blood volume stored in the veins of the skin. All of these reservoir functions can significantly affect mean systemic pressure, as their effective capacitance is altered, and blood is transferred to or from other portions of the vascular system.

The vasopressor hormone angiotensin II is implicated as a causative factor in many forms of hypertension. The renal sodium-retaining effects of angiotensin II are the primary mechanisms contributing to sustained blood pressure elevation, although the peptide has other significant vascular actions. Its effects on mean systemic pressure were analyzed in a series of studies in dogs in which angiotensin II was infused intravenously for 7 days, raising mean arterial blood pressure from the normal level of 100 to 160 mm Hg [5]. Blood volume remained unchanged, while mean systemic pressure rose from 9.5 to 12.6 mm Hg. The effect of the hormone was to increase the vascular tone, causing an increase in filling pressure at a constant blood volume. The increase in mean systemic pressure may have been caused by a decrease in the unstressed vascular volume and/or a decrease in capacitance of the system.

Right atrial pressure is normally approximately 0 mm Hg or atmospheric pressure. At a normal level of right atrial pressure, venous return will be normal as long as mean systemic pressure and resistance are normal. Guyton found that increasing right atrial pressure by 1 mm Hg decreased venous return by 14% in animals whose sympathetic nervous systems had been blocked [6]. Each additional 1 mm Hg increase resulted in a similar decrease in venous return, until atrial pressure reached 7 mm Hg, the mean systemic pressure, at which point flow into the heart ceased. The results of their study are plotted in Figure 2.2.

When right atrial pressure is reduced below the normal value of 0 mm Hg, a different venous return response pattern is observed. For the first 1 mm Hg reduction in right atrial pressure, venous return increases by 10%. But with subsequent 1 mm Hg increments in pressure reduction, the rate of rise in venous return falls progressively less until it reaches a steady level at pressures below −4 mm Hg. Further right atrial pressure reductions below −4 mm Hg will not increase venous return further. The negative right atrial pressure and venous return data are presented in Figure 2.3. The relationship becomes curved as pressure falls to approximately −2 to −3 mm Hg as the slope decreases progressively with additional reductions in atrial pressure. At approximately −4 mm Hg,

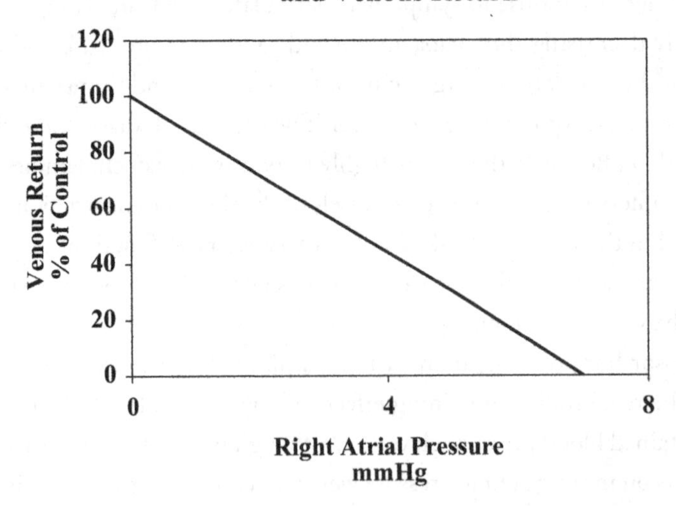

FIGURE 2.2: Relationship between right atrial pressure between 0 and 7 mm Hg and venous return. As atrial pressure is raised from the normal value of 0 to 7 mm Hg, venous return falls from the normal level to 0. The slope of the relationship is the inverse of the value of resistance to venous return. From reference [6].

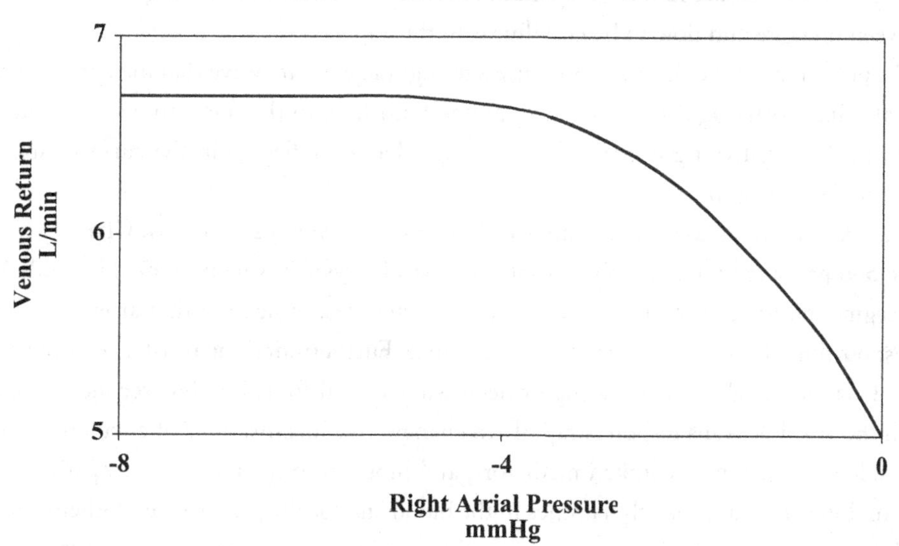

FIGURE 2.3: Relationship between right atrial pressure between 0 and −8 mm Hg and venous return. The relationship is curvilinear between −2 and −4 mm Hg due to progressively increasing resistance to venous return resulting from collapse of veins entering the thorax. Below −4 mm Hg, a plateau in venous return is reached. From reference [6].

the slope becomes 0, and further reductions do not cause additional increases in venous return. Below −4 mm Hg, venous return maintains a plateau at a flow rate of 20–30% above the normal value associated with right atrial pressure of 0 mm Hg.

The explanation for the nonlinear nature of the relationship in the negative pressure range of the right atrial pressure and the plateau below −4 mm Hg is the progressive collapse of veins as the luminal pressure falls below extramural pressure. Within the chest, the pressure averages approximately −4 mm Hg but cycles between values more negative during inspiration to slightly positive during expiration. As right atrial pressure, which is equal to venous pressure anywhere within the thorax, falls below atmospheric pressure, some veins just outside their point of entry into the thorax may collapse during inspiration, as their intraluminal pressure falls below atmospheric pressure. As central venous pressure falls lower, more veins may collapse for longer portions of the respiratory cycle, while below −4 mm Hg, essentially, all veins in the chest remain collapsed until the buildup of upstream blood increases their intraluminal pressure to −4 mm Hg or greater. The collapse of the veins increases resistance to venous return, which is the inverse of the slope of the relationship between flow and right atrial pressure. Ultimately, resistance becomes infinite below −4 mm Hg, preventing any increase in flow above that present at −4 mm Hg. The resistance increases progressively as right atrial pressure falls from approximately −2 to −4 mm Hg, causing the plotted relationship between pressure and flow to be curvilinear in this range.

The pulsations of the right atrium cause a retrograde pressure wave that may progress through the central veins to varying distances. These pulses contribute to the fluctuations in venous closure that occur in the negative right atrial pressure range that are reflected in the curve or splay of the pressure–flow relationship.

Changes in arterial as well as venous resistances affect venous return. In Chapter 1, the progressive blood pressure reductions throughout the vascular system were presented in Table 1.1. The greatest segmental pressure reduction occurs at the arterioles, indicating that arterioles contribute the largest portion of total systemic vascular resistance. Furthermore, the resistance of the arterioles is highly dynamic, capable of increasing or decreasing several folds in a few seconds. The smooth muscle in the arteriole walls responds rapidly to changes in concentrations of circulating vasoactive hormones, local metabolically linked mediators, and input from fibers of the sympathetic nervous system. Angiotensin II and catecholamines in the blood and locally produced endothelin are powerful arteriolar smooth muscle agonists, significantly affecting resistance to venous return.

In the experiment referred to above, in which angiotensin II was infused into dogs for 7 days, venous return remained unchanged while mean systemic pressure increased from 9.5 to 12.6 mm Hg. During this period, right atrial pressure increased slightly from 1.6 to 3.4 mm Hg. Calculating resistance to venous return during the control period from the pressure gradient for venous return (mean systemic pressure–right atrial pressure) and the rate of venous return (cardiac output) yields a value of 2.2 L/min/mm Hg (Figure 2.4). After 7 days of angiotensin infusion, resistance to venous return increased to 3.3 L/min/mm Hg, a 50% increase resulting from the constriction of arterioles and possibly of portions of the venous system as well.

Locally produced and circulating nitric oxide, prostacyclin, and prostaglandin E_2 are vascular smooth muscle antagonists, producing arteriolar dilation and reduction of resistance to venous return. Local tissue metabolism, in particular, aerobic metabolism, strongly affects arteriolar resistance. Activity that reduces tissue pO_2 especially elicits significant arteriolar dilation and reduction in resistance to venous return.

The linkage between total body tissue oxygen demand and resistance to venous return is a fundamental mechanism governing control of cardiac output. This is the basic mechanism by which the cardiovascular system responds to changes in demand for cardiac output as metabolic rate changes. Other means of cardiovascular control may take part in responses to metabolic changes, but this connection of tissue oxygen demand to resistance to venous return is of overriding significance. Oxygen demand is a strong determinant of resistance to venous return over periods ranging from seconds to hours and in long-term and steady-state conditions. If demand is elevated for

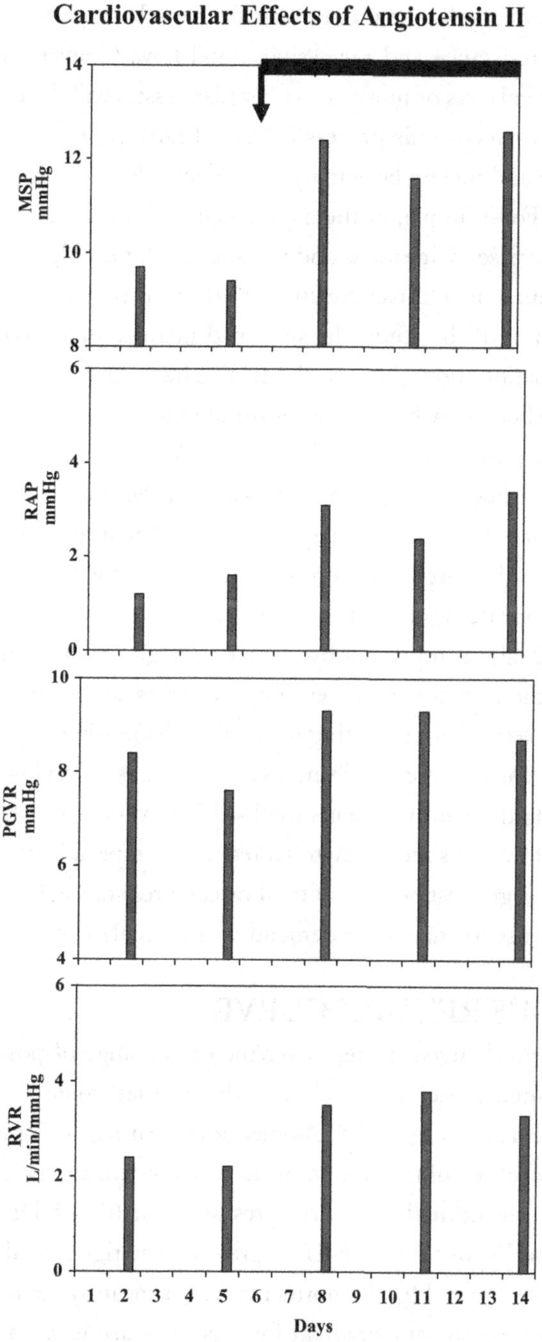

Cardiovascular Effects of Angiotensin II

FIGURE 2.4: Cardiovascular effects of angiotensin II. In a study on dogs, after a 7-day control period, angiotensin II was infused intravenously for an additional 7 days. Values of mean systemic pressure (MSP), right atrial pressure (RAP), pressure gradient for venous return (PGVR), and resistance to venous return (RVR) are presented. From reference [5].

extended periods of days or weeks, new microvascular vessels grow through the tissue in need, decreasing local vascular resistance and increasing blood flow. Conversely, if blood flow exceeds demand for periods of several days or more, microvascular vessels will degenerate, reducing vascular density and increasing resistance. This process is termed rarifaction and normally normally takes place in tissues whose use and metabolic activity are reduced. Rarifaction also may occur if arterial blood pressure increases. For example, in the angiotensin II infusion experiment, the infusion resulted in a steady-state increase in arterial blood pressure of 60 mm Hg by affecting renal function, and the peptide had an immediate direct constrictor effect on the arterioles throughout the body. But during the 7-day course of the study, the sustained increase in arterial pressure may have induced microvascular rarifaction throughout the body. The immediate and delayed increases in tissue resistance throughout the body may both have contributed to the increase in observed resistance to venous return during the infusion period.

Venous resistance makes up about 15% of total vascular resistance and is not regulated as actively as arterial resistance. The relatively large diameter of central veins presents little resistance to flowing blood, although they are easily compressed and flattened by surrounding tissue. When they are compressed, they create significant resistance. For example, many veins entering the thorax over the first ribs are partially compressed by the sharp angle of the path over the bone. In the abdomen, the weight of the viscera may flatten the great veins, and in the neck, atmospheric pressure prevents the jugular veins from assuming a rounded shape when a person is upright. Within the thorax, the veins may collapse if central venous pressure falls much lower than the atmospheric pressure. Even considering these impediments to blood flow, venous resistance is a relatively minor component of resistance to venous return. Arterial resistance, especially that portion resulting from the arterioles, makes up the greatest portion of total vascular resistance. It is this portion that is most actively regulated in response to changes in demand of the circulatory system.

2.2 THE VENOUS RETURN CURVE

If right atrial pressure were changed in steps over the entire range of possible atrial pressures and venous return were measured at each point, plotting the data set would yield a complete venous return curve, which is presented in Figure 2.5. As mentioned earlier, such measurements would have to be made during total blockade of the autonomic nervous system so that circulatory reflexes would be normal. Notice that, at the normal right atrial pressure value (0 mm Hg), venous return is 100%, which is 5 L/min in man. Venous return falls progressively as right atrial pressure increases, until right atrial pressure reaches 7 mm Hg, the normal value for mean systemic pressure. At that point, venous return is 0 because the pressure gradient for venous return is 0. As right atrial pressure falls below 0, the venous return curve increases at a progressively declining rate until flow reaches a plateau at approximately −4 mm Hg. As discussed above, the reason for the curvilinear nature in this portion

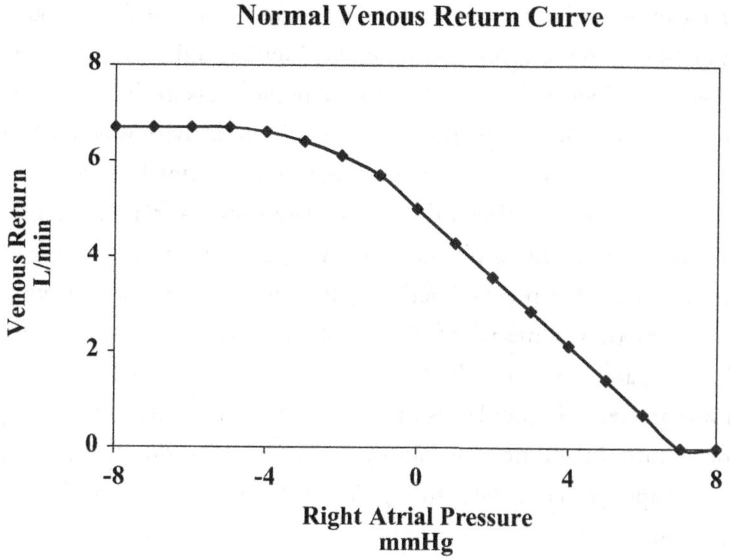

Normal Venous Return Curve

Venous Return
L/min

Right Atrial Pressure
mmHg

FIGURE 2.5: The complete venous return curve over the range of right atrial pressure from −8 to 8 mm Hg. Venous return values are for humans.

of the relationship, termed the *transition zone*, is the progressive increase in vascular resistance due to the collapse of increasing numbers of veins as right atrial pressure becomes more negative.

Such a function curve can reveal important characteristics of the circulation. First, the value of cardiac output or venous return at a given level of right atrial pressure can be read directly from the curve. Similarly, the value of mean systemic pressure is easily determined from the value of the *x*-axis intercept. For a given level of right atrial pressure, the pressure gradient for venous return can be calculated from the difference between the mean systemic pressure and the value of right atrial pressure. The resistance to venous return can also be calculated from the pressure gradient for venous return and rate of venous return at any level of right atrial pressure. Finally, the lower limit of right atrial pressure that will affect venous return, the plateau pressure, can be determined from inspection of the graph. These circulatory characteristics are key elements in understanding the regulation of cardiac output.

2.3 ALTERATIONS OF THE VENOUS RETURN CURVE

The characteristics of the venous return curve can be altered dramatically within seconds by rapidly acting physiological mechanisms and for indefinitely extended periods by long-acting responses of the circulatory control system. The curve can also be affected by many pathological circumstances. However, all factors modifying the curve act by altering one or more of its basic characteristic: mean systemic pressure, plateau, slope, and right atrial pressure.

Changes in mean systemic pressure resulting from alteration of either blood volume or vascular tone cause parallel shifts in the venous return curve. Guyton and associates extensively analyzed the relationship between blood volume and mean systemic pressure in anesthetized, open-chest canine preparations in which the sympathetic nervous system activity was blocked by total spinal anesthesia, but a constant level of sympathetic tone was maintained by continuous intravenous infusion of very small amounts of epinephrine [7, 8]. Their results from one animal are presented in Figure 2.6. By removing or adding blood, mean systemic pressure was changed over the range 4.7–10.6 mm Hg, and other data points describing the venous return curves were obtained.

The increase in blood volume raised mean systemic pressure and shifted the venous return curve to the right in a parallel manner. Notice that, at each level of right atrial pressure, the rate of venous return was greater at higher levels of mean systemic pressure, due to the greater pressure gradient for venous return. In addition, previously in this presentation, the venous pressure plateau was stated to begin at approximately −4 mm Hg; in this figure, it is at 0 mm Hg because it was an open-chest preparation.

When blood volume is increased very rapidly even in areflexic preparations, the immediate change in mean systemic pressure begins to wane almost immediately. The effect is the result of stress relaxation of the walls of the large capacitance vessels, although fluid transudation out of the vascular space into the interstitium undoubtedly contributes gradually to the phenomenon as well. In one study, Guyton and associates rapidly infused 35% of the blood volume of a dog and then

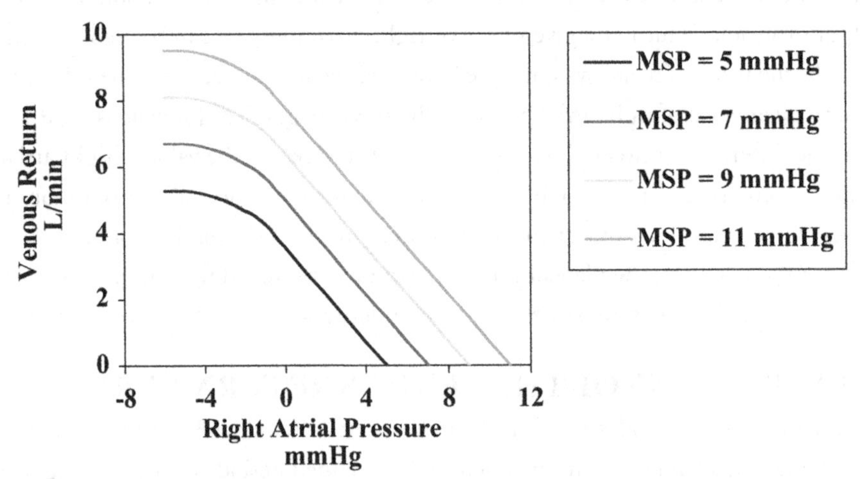

FIGURE 2.6: Effect on the venous return curve of changes in mean systemic pressure resulting from alterations in blood volume from one dog. The anesthetized animal had all autonomic reflexes blocked. Changes in mean systemic pressure resulted in parallel shifts in the curve. From references [7] and [8].

followed mean systemic pressure over the following minutes [9]. Mean systemic pressure reached its maximum (24 mm Hg) at the conclusion of the 1-minute infusion and then began immediately to decline asymptotically toward a steady-state value somewhat above the initial level with an estimated half-time of 2–4 min. Their estimate of the time course of change in mean systemic pressure following a large infusion is presented in Figure 2.7.

They reported similar findings in the opposite direction following step decreases in blood volume, although possibly with longer response times. In either case, the effect of volume change on cardiac output is consistent with the changes in pressure gradient for venous return resulting from the change induced in mean systemic pressure.

Changes in vasomotor tone affect mean systemic pressure by altering vascular capacitance and unstressed vascular volume. Guyton's laboratory analyzed the quantitative effects of vasomotor tone on mean systemic pressure and venous return in anesthetized animals prepared with total spinal anesthesia and, therefore, with all autonomic vasomotor reflexes abrogated. To simulate increasing levels of vasomotor activity, epinephrine was infused intravenously at a rate of up to 3.5 μg/kg/min [4]. With all reflexes blocked and no epinephrine infusion, mean systemic pressure fell from 7 to 5 mm Hg and arterial pressure to 41 mm Hg. Increasing epinephrine infusion to a maximal level [3.5 μg/kg/min) raised mean systemic pressure to 19 mm Hg and blood pressure to 184 mm Hg.

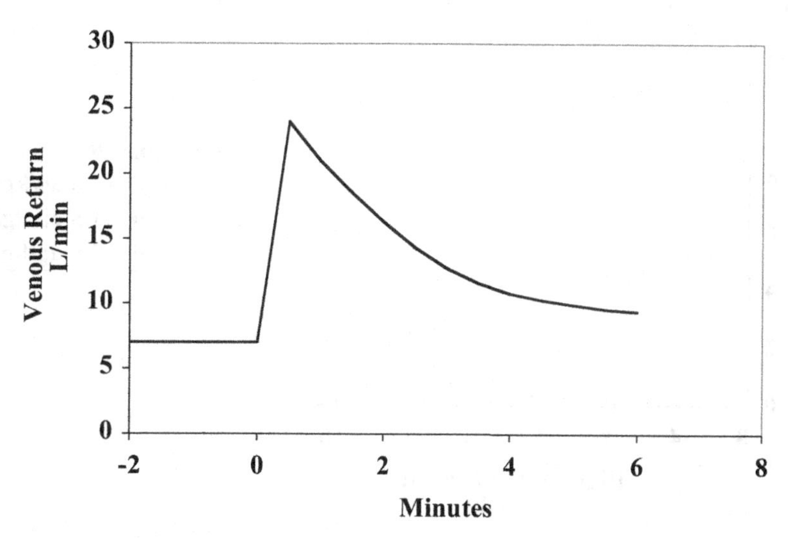

FIGURE 2.7: Effect of vascular stress relaxation on the value of mean systemic pressure following a large infusion of blood into an anesthetized areflexic dog. The half-time of the response appears to be between 2 and 4 min. From reference [9].

They obtained data for four venous return curves across this range of epinephrine infusion, which are illustrated in Figure 2.8.

The effect of increasing autonomic reflex replacement was to elevate mean systemic pressure and shift the venous return curve upward and to the right in an approximately parallel manner, similar to that described for volume increases. The range of simulated vasomotor activity in theses studies appears to span the range of naturally occurring activity, as judged by comparisons with mean systemic pressure and arterial blood pressure during several manipulations known to be associated with stimulation of vasomotor reflexes [4, 10, 11]: eliciting the maximal carotid sinus reflex increases mean systemic pressure to about 10 mm Hg; the Cushing reflex, which is believed to elicit the most intense naturally occurring sympathetic nervous system reflexes, raises mean systemic pressure to 17 mm Hg; and infusion of a maximal dose of norepinephrine increases mean systemic pressure to 15 mm Hg.

Changes in conditions outside the vascular system can affect mean systemic pressure and venous return. In particular, factors that compress portions of the vascular system, thereby decreasing capacitance, will increase mean systemic pressure.

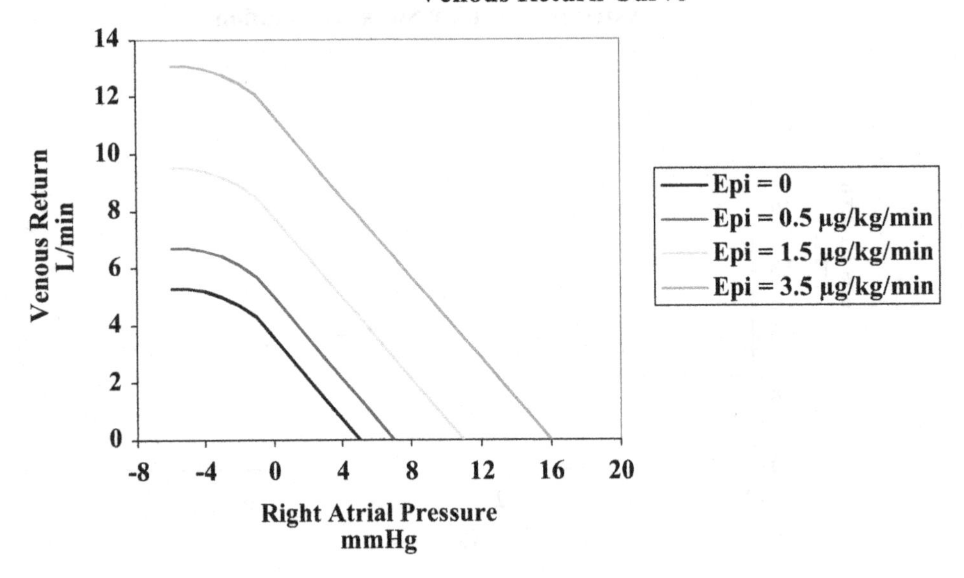

FIGURE 2.8: Shifts in venous return curves caused by varying epinephrine infusion rate in areflexic dogs. The rates of infusion were designed to mimic the effects of the complete range of autonomic nervous system activity on the vascular system. From reference [4].

Abdominal compression can occur during nearly any whole body physical exercise, such as walking or running, athletic events of all types, weight bearing exercises, and movements that require balance. The "core" muscles are activated in all movements involving simultaneous movement of several body parts, and the abdominal muscles make up a large portion of the core muscles. Contraction of the musculature surrounding the abdominal contents increases pressure throughout the abdomen, including the pressure around the large veins, the liver, spleen, and other structures having large capacitance. Consequently, blood is transferred from these structures into the rest of the circulation, raising mean systemic pressure. Compression of the abdomen with the hands can double mean systemic pressure in anesthetized dogs [10], and the effect of forceful contraction of all the muscles that make up the walls of the abdominal cavity may raise the pressure even higher. The effects of increasing abdominal pressure on mean systemic pressure are nearly immediate, occurring in within 1 s.

Positive pressure breathing increases pressure throughout the thorax, decreasing the blood volume contained in the vascular structures within it. The volume of the heart and pulmonary veins as well as the blood vessels of the lungs all decrease during positive pressure breathing. In addition, the muscular work required to breath against positive pressure requires forceful involvement of the abdominal muscles. Together, these effects can at least double the mean systemic pressure [10].

Contraction of any muscle increases the pressure within the sheath of connective tissue surrounding it. This forces blood from the muscle into the rest of the circulation. When many large muscles contract at the same time, as during standing from a sitting position, running, weight-bearing exercise, or any action requiring coordinated movement of several body parts, the movement of blood volume out of the muscles can significantly raise mean systemic pressure. Electrical stimulation of muscles of the hind legs and abdomen can increase mean systemic pressure to as high as 30 mm Hg within 1 s of the start of contraction [12].

Peripheral edema can increase interstitial fluid pressure to as high as 10 mm Hg, which compresses peripheral veins and elevates mean systemic pressure [4]. Similarly, ascites fluid accumulation in the abdominal cavity can increase pressure on the capacitance structures within the abdomen.

Increasing resistance to venous return decreases the slope of the venous return curve. Although an increase in resistance alone will not alter the mean systemic pressure, venous return will be reduced at each level of right atrial pressure, and the plateau value will be decreased. Reducing resistance to venous return will have the opposite effects on the curve. Examples of venous return curves at normal resistance (100%), at 50%, and 200% are illustrated in Figure 2.9.

Resistance to venous return can be manipulated experimentally by opening and closing a large shunt connecting the arterial and venous systems. The venous return curves for a group of dogs prepared with such a shunt are presented in Figure 2.10 [13].

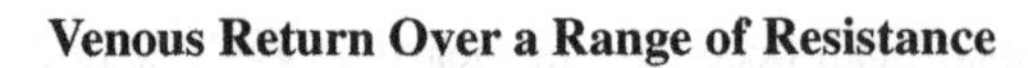

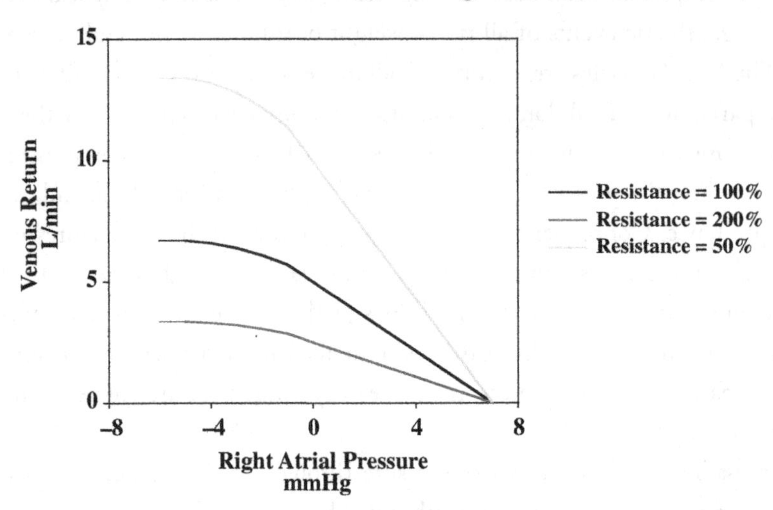

FIGURE 2.9: Effect on the venous return curve of altering resistance to venous return.

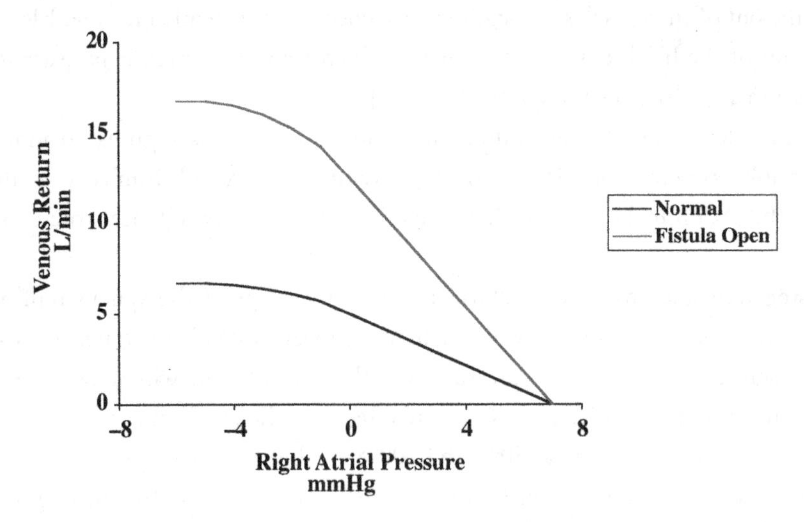

FIGURE 2.10: Shift in the venous return curve associated with opening a large arterial to venous shunt. From reference [13].

The mean systemic pressure was approximately the same in both conditions, although when the shunt was opened and resistance to venous return was sharply reduced, the slope was much greater. Venous return also increased at each level of right atrial pressure, and the plateau was elevated.

The quantitative effects on venous return of changes in venous resistance are much greater than proportionally similar changes in arterial resistance. Ohm's law can be useful in understanding flow of fluid through rigid tubes, but because the circulatory system is made up of compliant vessels, modification of the law can be made in order to make it more useful in analysis of flow through the cardiovascular system. Instead of using one simple term for systemic resistance, arterial and venous resistances can be weighted separately according to the capacitances of the two segments of the vascular system. The importance of venous resistance is even more striking when viewed in light of the experimental data, indicating that the ratio of venous to arterial capacitance may be as large as 18:1 [3]. The derivation of the modified equation is presented in Chapter 4.

2.4 SUMMARY

Venous return and, consequently, cardiac output are functions of the pressure gradient for venous return and the sum of the resistances of the arterial and venous segments. The pressure gradient is affected by factors that increase or decrease mean systemic pressure and/or right atrial pressure. Resistance to venous return is affected by factors that cause changes in smooth muscle tone of resistance vessels or changes in pressure in the tissue surrounding thin-walled venous structures.

The implications of these statements are broad and significant and include the following:

- Cardiac output is very sensitive to the pressure gradient for venous return. An increase in mean systemic pressure of only a few mm Hg, such as those occurring in muscular activity or with an increase in blood volume, will result in immediate increases in cardiac output. Conversely, even small increases in right atrial pressure of a few mm Hg, such as those occurring in acute heart failure following myocardial infarction, will result in significant reductions in cardiac output. Furthermore, the most that right atrial pressure can possibly increase is to a level approaching mean systemic pressure.
- Cardiac output is affected by systemic vascular resistance, which is the sum of capacitance-weighted arterial and venous resistances, with an increase in total systemic resistance resulting in a reduction in cardiac output. Changes in venous resistance will have a much greater effect on total systemic resistance than equivalent percentage changes in arterial resistance.

• • • •

CHAPTER 3
Cardiac Function

The heart is a powerful, complex organ that has fascinated scholars, poets, and physicians for centuries. It provides the energy for the flow of life-giving blood through the circulatory system, and its strong, rhythmic beating is the very symbol of life. Cardiac physiology has been studied intensively, yielding an advanced understanding of the heart's intrinsic function. But for this presentation, the mechanisms of the heart's inner workings are not the subject of greatest interest. Here, the object is to provide an analysis of regulation of cardiac output, and to that end, what is required is knowledge concerning how the heart functions as a component of the circulatory system. The healthy heart can increase its output approximately 3- to 4-fold, but only if blood flows into the right atrium at three or four times the normal rate. If the heart pumps blood into the aorta or pulmonary artery faster than it enters the atria, pressure in the atria will fall, the veins will begin to collapse, resistance to venous return will increase, and flow into the atria will fall, thereby limiting cardiac output to the rate at which blood returns to the heart. Regardless of how forcefully the myocardium contracts, the flow of blood from the heart cannot exceed the rate of venous return for longer than a few seconds. In conditions of health, the function of the heart alone cannot determine cardiac output, for the heart is only one component of a complex system. Therefore, the vitally important and fascinating subjects concerning the physiology of the myocardium will not be included here; rather, the focus will be on the characteristics of the heart as a pump within the system that regulates cardiac output.

3.1 CARDIAC OUTPUT CURVES

One of the first observations made by pioneering cardiac investigators was that the function of the heart was strongly influenced by the level of atrial pressure, the force of contraction increasing dramatically as atrial pressure increased. Therefore, cardiac performance has been expressed as various functions of right or left atrial pressure for nearly a century, beginning with Patterson and Starling [14] and Wiggers [15]. Atrial pressure may be plotted as the independent variable with various measures of cardiac function plotted as the dependent variable. Four types of function curves have been used most frequently: cardiac output plotted versus mean right atrial pressure; pressure output of each ventricle plotted versus ventricular end diastolic volume; ventricular work output plotted versus mean atrial pressure; and ventricular power output plotted versus mean atrial pressure. Each of these conveys valuable amounts of information of different types. However, the function curve relating cardiac output to mean right atrial pressure is most useful for analysis of the heart's function in the circulatory system in which venous return and the cardiac pump work together in series.

For nearly all circumstances that will be considered here, the functions of the right and left ventricles can be considered as one unit. The main exceptions are conditions in which the right or left ventricles undergo pathological changes separately or when other factors create imbalance between the pumping abilities of the two ventricles. But in other conditions, the heart can be considered accurately as a single unit in which cardiac output is related to right atrial pressure.

Cardiac function curves can be derived from experiments in rats or larger experimental animals in which right atrial pressure is changed in a controlled manner while cardiac output is measured continuously. To obtain data from closed-chest preparations, right atrial pressure may be controlled by rapidly infusing or withdrawing blood through a large bore catheter placed in the right atrium from the external jugular vein. Ideally, the data for the complete curve should be obtained within 45–60 s, before the conditions of the heart and circulatory system change in response to experimental manipulation. If the experiment is conducted properly, data points for a function curve similar to the curve in Figure 3.1 will be obtained. At the normal right atrial pressure (0 mm Hg), cardiac output will be at its normal value, approximately 5 L/min for a normal size healthy human. As right atrial pressure is reduced below 0 mm Hg, cardiac output decreases until, in the normal, closed-chest situation, at right atrial pressure of −2 to −3 mm Hg, cardiac output reaches 0. The progressive reduction in output is due to the Starling effect (reduction in length of stretch of cardiac muscle fibers, discussed in the following paragraph). Increasing right atrial pressure above normal is associated with very steep increases in cardiac output, until at a right atrial pressure of 3–4 mm Hg, cardiac output increases to its maximum, approximately three to four times the normal level. Further increases in right atrial pressure are not associated with additional increases in cardiac output, a plateau level being reached at right atrial pressure of 3–4 mm Hg.

Most noteworthy is the large increase in cardiac output with each small incremental increase in the right atrial pressure in the range between −1 and 1 mm Hg. This reflects the heart's capacity to respond robustly and rapidly to even small changes in atrial pressure within the normal range of right atrial pressure. This exceptional capability is due to the effect of stretch on the cardiac muscle fibers. With increasing ventricular end diastolic pressure, ventricular volume increases, lengthening the cardiac muscle fibers. Consequently, the force of contraction of the fibers increases so that the pumping ability of the ventricle is markedly augmented. The effect was described by Starling and associates nearly a century ago and is referred to as Starling's law of the heart [14, 16]. This intrinsic cardiac function enables the heart to double its output within a single contraction when end dia-

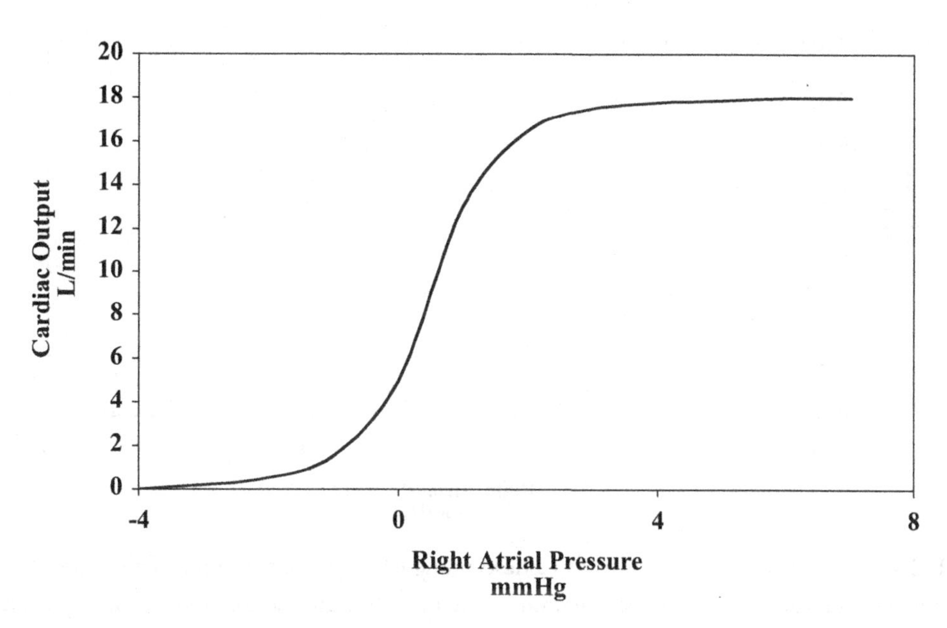

FIGURE 3.1: The cardiac function curve, with cardiac output values scaled to normal human levels.

stolic pressure is increased from normal to 1 or 2 mm Hg. However, the heart reaches its limit of force generation when the muscle fibers are stretched to their greatest effective length. The mean ventricular end diastolic pressure at which the cardiac output reaches a plateau is the filling pressure that produces the greatest force generation from the ventricle.

If the condition of the heart deteriorates during the data collection period, the resulting function curve may have a "hump" and a descending limb at high levels of right atrial pressure. Prolonging the experimental procedure to obtain data for longer than several minutes may fatigue or damage the heart, especially at high levels of right atrial pressure. Under such conditions, the data obtained may not yield a curve with a stable plateau. Instead, it may have a hump and a descending leg at higher levels of right atrial pressure, as in the one plotted in dashes in Figure 3.2.

In reality, the curve represents a compilation of data from a series of time-dependent function curves, each reflecting the progressive weakening of the heart during the course of the study. As the experiment progresses to higher levels of right atrial pressure, the myocardium weakens so that the data collected were not from the heart in its initial condition. Consequently, at each level of deterioration, the function curve is shifted downward. In reality, the data from an excessively lengthy study could be described by a series of time-dependent function curves such as those presented in Figure 3.3.

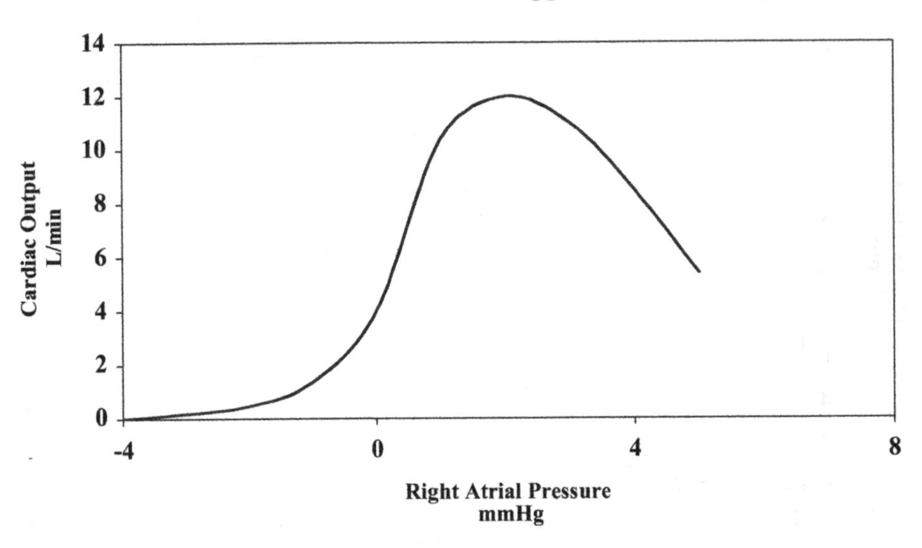

FIGURE 3.2: A cardiac function curve with a descending leg in the upper range of atrial pressure. The descending portion may be due to deterioration of the myocardium due to fatigue or injury associated with excessive length of time required for data collection.

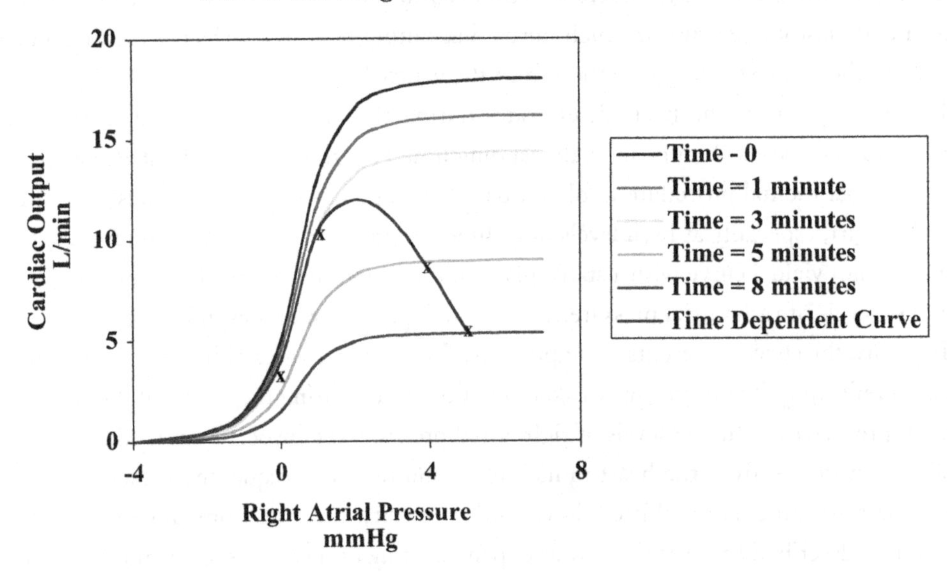

FIGURE 3.3: A series of function curves obtained over an extended period illustrating how a descending portion could be erroneously obtained.

If an assumption is made incorrectly that the heart's function remains unchanged during the time the right atrial pressure is raised, the data collected over the 4-min period would be plotted as a series of data for one curve, with a descending leg at higher levels of right atrial pressure. Such curves have appeared in the literature as "classic" cardiac function curves, although in reality, they are data points obtained from a series of time-dependent curves from the progressively weakening heart. Within the normal range of right atrial pressure, the accurate cardiac function curve has a stable plateau [17–19].

3.2 FACTORS THAT ALTER CARDIAC OUTPUT BY CHANGING THE EFFECTIVENESS OF THE PUMPING ABILITY OF THE HEART

Systemic resistance and arterial pressure affect the cardiac function curve. Data for cardiac function curves may be obtained from experiments in which either the systemic resistance or the systemic arterial pressure is controlled. If the resistive load is held constant, arterial pressure will increase progressively as cardiac output increases, and consequently, the effectiveness of the pumping ability of the left ventricle will be limited as the outflow pressure increases to higher levels, as illustrated in Figure 3.4.

If the load is reduced, cardiac output will be affected only slightly. This is because the output of the right ventricle is near maximal even under conditions of normal arterial resistance, so that reduction in arterial load will not result in a significant increase in output of the right ventricle or the left ventricle.

But if the data are obtained from a constant pressure study rather than one in which resistive load is held constant, the pumping ability of the heart will not be constrained by continually increasing arterial pressure as cardiac output rises. Figure 3.5 presents the expected shapes for function curves when resistive load is held constant and when pressure load is maintained at a fixed, normal level.

Most data for cardiac output function curves are obtained from intact, closed-chest animal preparations in which neither arterial pressure nor resistive load can be controlled. But in a properly conducted study, during which the data are rapidly collected, resistance changes very little and may even decrease as the increasing arterial pressure stretches the arteries slightly. These studies cannot be termed constant resistance but can be called "normal resistance" function studies. Most of the cardiac output function curves in this presentation were derived from data obtained with normal resistive load.

Sagawa and associates performed an in-depth analysis of the relationships between left atrial pressure and cardiac output [21] and between right atrial pressure and cardiac output [22], in both

Effect of Changes in Fixed Resistive Load on Cardiac Function

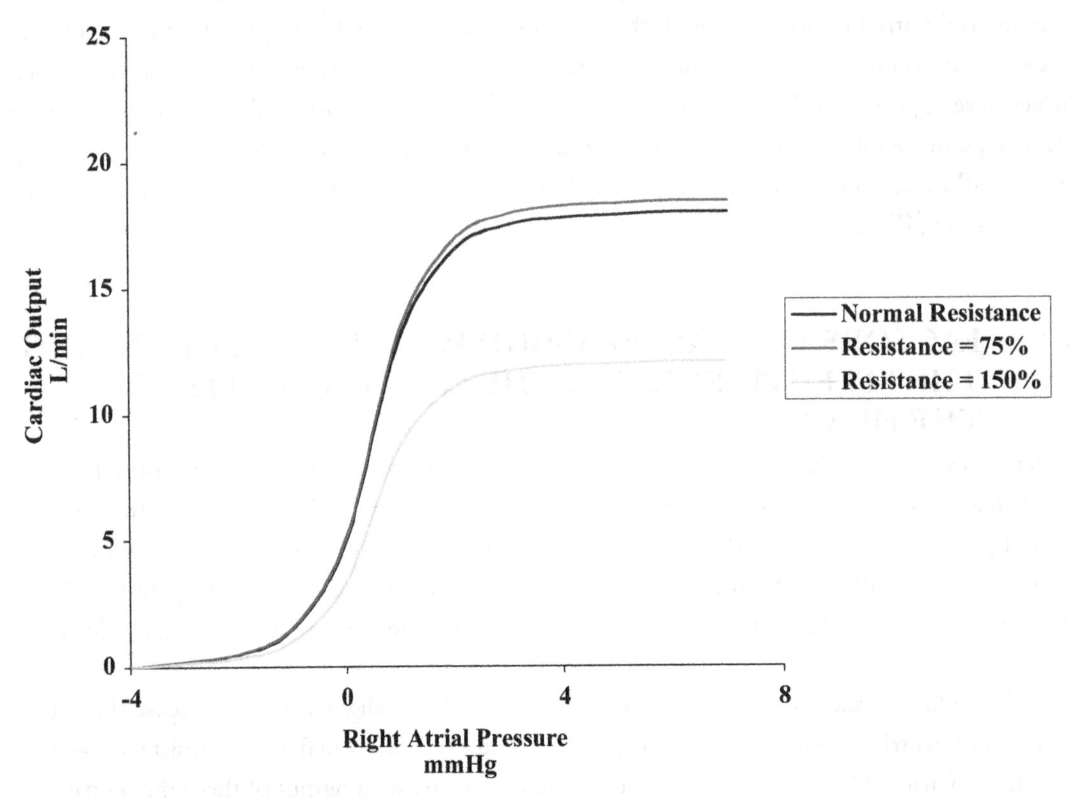

FIGURE 3.4: Cardiac function curves obtained at different fixed levels of arterial resistance. Reducing resistance to 75% of normal results in only a slight increase in the slope and plateau, although increasing resistance to 150% of normal significantly limits output over the higher range of atrial pressure.

cases with arterial pressure controlled in steps over a range of 0–250 mm Hg. Their data are plotted in three dimensions in Figure 3.6.

The output of the left ventricle increases progressively as left atrial pressure increases to as high as 20 mm Hg (3.6 A), especially when mean arterial pressure is controlled at levels below 100 mm Hg. Little evidence of a stable plateau value appears until arterial pressure is maintained near 200 mm Hg. In contrast, the output of the total heart (3.6 B) fails to increase as right atrial pressure is raised above 8 mm Hg (open-chest preparation) even when arterial pressure is maintained at 100 mm Hg. Taken together, these observations imply that the pumping ability of the heart is limited by the ability of the right ventricle to increase its output to levels greater than those attained at relatively low right atrial pressure and that cardiac output is relatively insensitive to the effects changes in mean arterial pressure below 150 mm Hg.

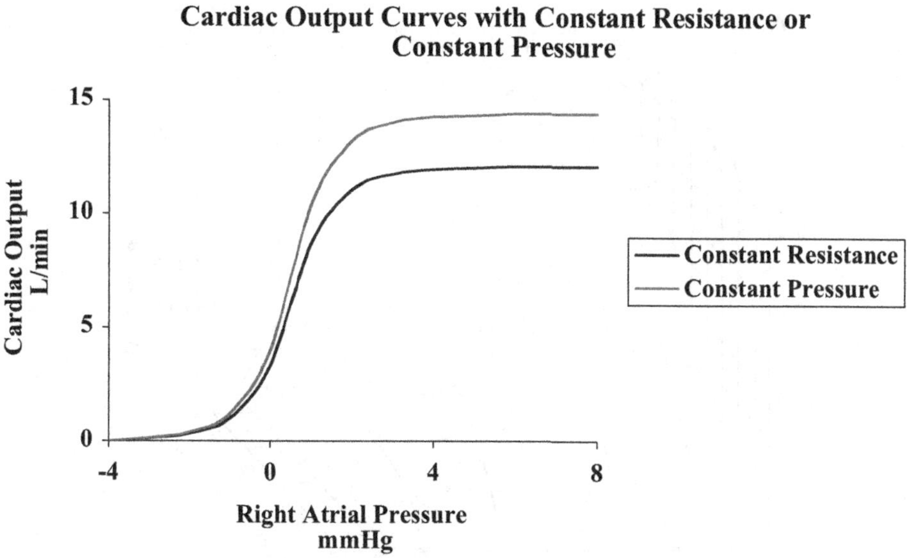

Cardiac Output Curves with Constant Resistance or Constant Pressure

Right Atrial Pressure (mmHg)	Constant Resistance (L/min)	Constant Pressure (L/min)
-4	0	0
0	~3	~3.5
4	~12	~14
8	~12	~14.3

FIGURE 3.5: Differing effects of a fixed resistive load and a fixed pressure load on the cardiac function curve.

Increasing heart rate up to the optimal level increases the slope and plateau value of the cardiac function curve. Changing only the heart rate affects the pumping ability of the heart, even when all other factors remain constant. With increases in rate, the slope increases as well as the plateau until the rate is approximately 80–100% greater than normal (Figure 3.7). At some rate less than the maximal heart rate, the greatest increase in pumping ability is achieved; above this optimal rate, cardiac output declines progressively. The decease in function at supraoptimal heart rate is due to inadequate time for filling of the heart during diastole. Possibly, the pressure gradient for flow into the ventricles may influence the optimal heart rate, with higher levels of pressure gradient being associated with higher optimal heart rates.

The autonomic nervous system affects both heart rate and strength of contraction. Cardiac output must be able to respond quickly to the needs of the body in response to a variety of normal changes in activity or environment. The autonomic nervous system can produce rapid changes in cardiac function, with the sympathetic system increasing and the parasympathetic decreasing the pumping ability of the heart.

The sympathetic system acts via the adrenergic cardiac nerves that innervate the atria and ventricles and release norepinepherine, and by secretion of epinephrine and norepinepherine from the adrenal medulla, reaching the heart through the circulation. In both cases, the effect on the heart is an increase in heart rate and in cardiac contractility, shifting the cardiac function curve to

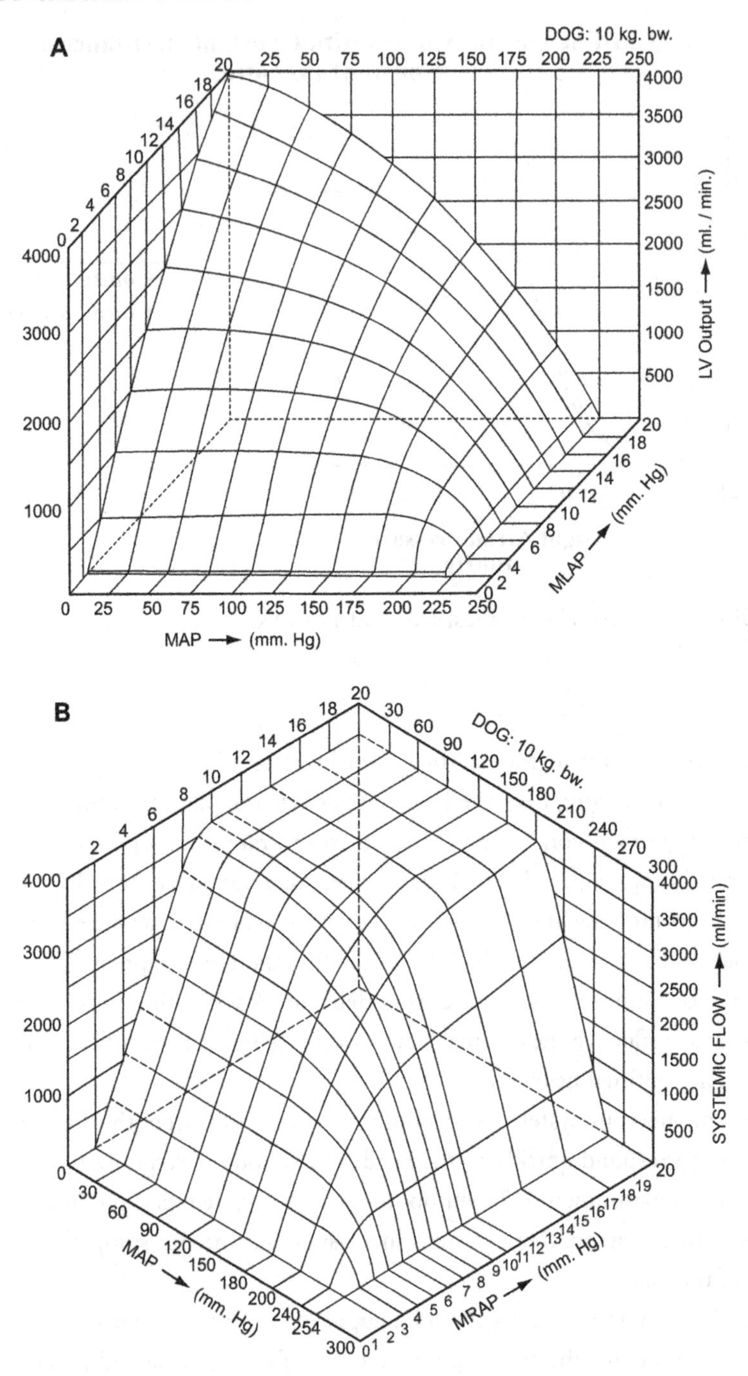

FIGURE 3.6: The three-dimensional interrelationship between atrial pressure, arterial pressure, and cardiac output: (A) left atrial pressure–arterial pressure cardiac output function; (B) right atrial pressure–arterial pressure cardiac out function. Atrial pressure and arterial pressure were varied independently at controlled values while cardiac output was measured continuously. Reproduced with permission of Elsevier from reference [20], based on data from references [21] and [22].

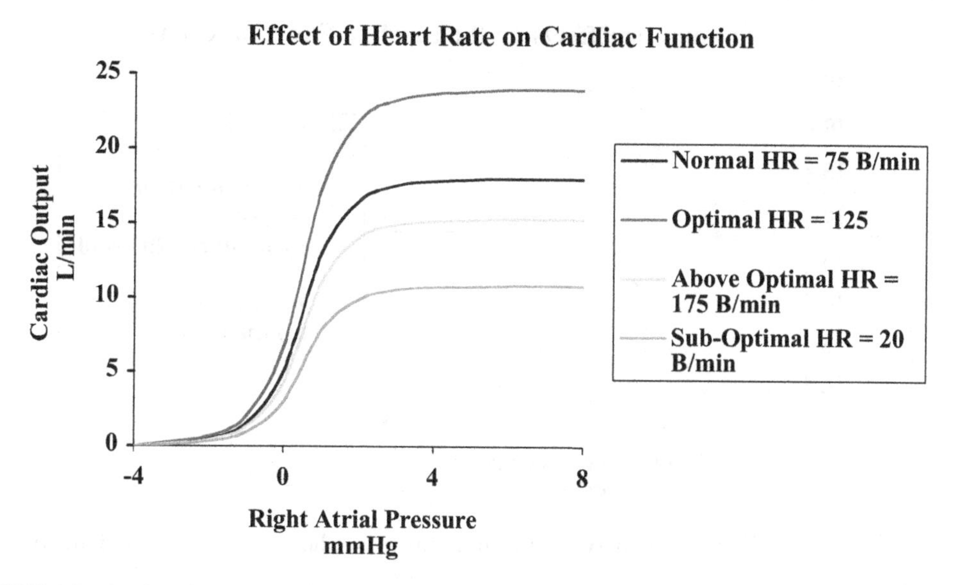

FIGURE 3.7: Cardiac function curves obtained at different fixed heart rates.

the left due to the effect on the slope of the curve, and increasing the plateau value. Interrupting sympathetic tone to a resting animal will reduce heart rate by only a few beats per minute, demonstrating that, in the normal resting condition, sympathetic stimulation of the heart is minimal. But if the sympathetic cardiac nerves are maximally stimulated while vagal input is blocked, heart rate can more than double. The upper curve in Figure 3.8 illustrates the shift in the function curve from its normal position to the maximal sympathetic stimulated state. With maximal stimulation, the plateau is increased to approximately 70% above normal [18]. If, in a normal subject or animal, all sympathetic nervous system activity were blocked, the cardiac function curve would be shifted downward and to the right due to withdrawal of the normal, tonic sympathetic stimulation.

The data used for Figure 3.8 were obtained from studies in which right atrial pressure was controlled as described previously to obtain complete function curve data. However, in an intact animal preparation or in a healthy human, if sympathetic stimulation to only the heart increases strongly, cardiac output may increase slightly or may remain unchanged while right atrial pressure will decrease. This combination may appear paradoxical, but it can be explained by the adrenergic augmentation of myocardial contractility increasing the slope of the function curve, shifting it to the left; however, because venous return does not rise, cardiac output will remain nearly unchanged. With adrenergic stimulation, the heart is operating on a shifted curve, one at which the nearly unchanged cardiac output is achieved at a reduced level of right atrial pressure. The curves describing the two conditions and the change in right atrial pressure at a constant cardiac output are illustrated in Figure 3.9.

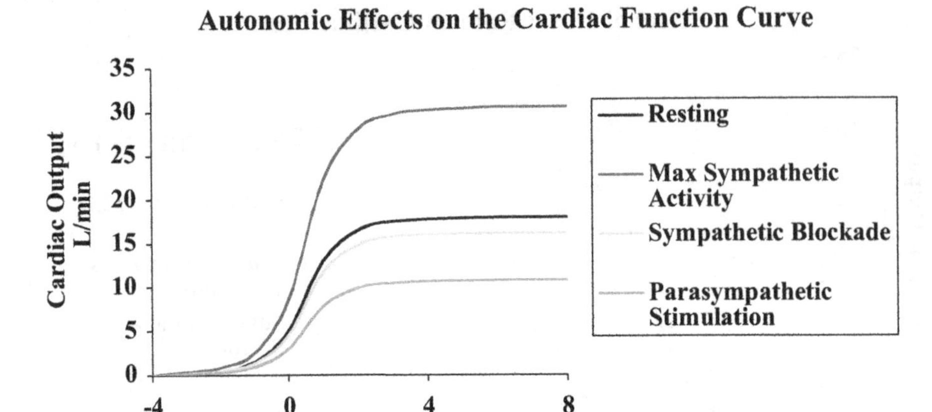

FIGURE 3.8: Effects of autonomic nervous system input to the heart on the cardiac function curve. From reference [18].

Parasympathetic input to the heart is mediated by acetylcholine released from the branches of the vagus nerve that innervate the sinoatrial node in the right atrium. Some cholinergic fibers originating from the vagus are also distributed to the ventricles. Increases in parasympathetic activity decrease heart rate from the normal level and can actually stop the heart for a few seconds if stimulation is maximal, although the heart escapes and rate will stabilize at approximately one-third

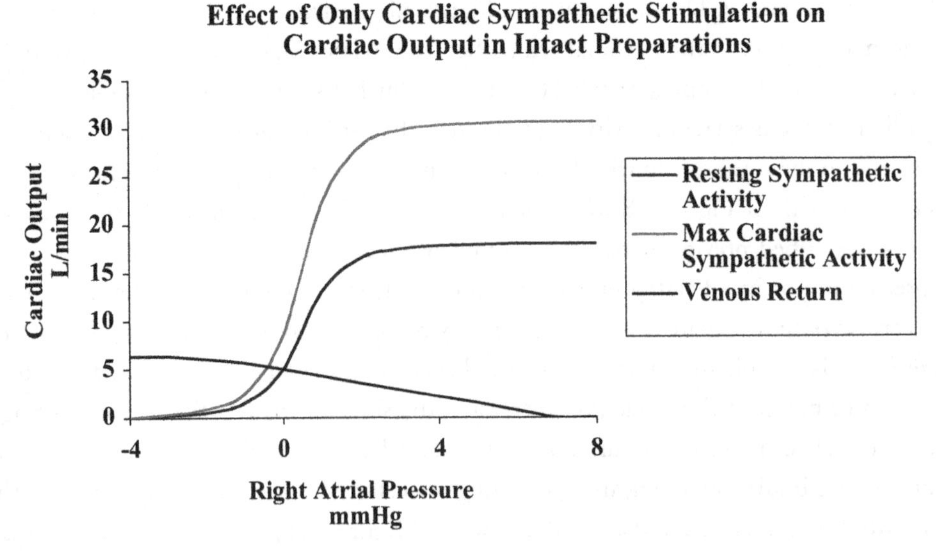

FIGURE 3.9: Effect of maximal cardiac sympathetic stimulation on the cardiac function curve in an intact preparation.

the normal rate. Totally blocking vagal input while an animal is in a resting state will increase heart rate by approximately 35%. Parasympathetic activity has some negative effect on ventricular contractility as well [23, 24], although the effects are not nearly as significant as the stimulatory effects of sympathetic stimulation. Figure 3.8 illustrates the shift in the cardiac function curve resulting from maximal parasympathetic activity after escape (blue curve).

Heart rate is determined by a complex interaction between the effects of sympathetic and parasympathetic systems. Figure 3.10, derived from data of Levy and Zeiske [26], illustrates the heart rates resulting from combined frequencies of stimulation of the parasympathetic and sympathetic nerves innervating the heart. The resulting three-dimensional surface reveals the nature of the complexity. Notice that, as vagal stimulation increases, the effect of sympathetic stimulation on heart rate decreases, so that at the highest vagal stimulation (8 Hz), the heart rate response to sympathetic stimulation is minimal, even at the highest rate of sympathetic stimulation. At lower vagal frequencies, the response to elevated sympathetic discharge rates is several times greater. A similar relationship exists between the changes in heart rate in response to vagal stimulation at fixed levels of sympathetic tone, although the effect of parasympathetic innervation is not attenuated to the extent observed with the sympathetic response at high levels of vagal input.

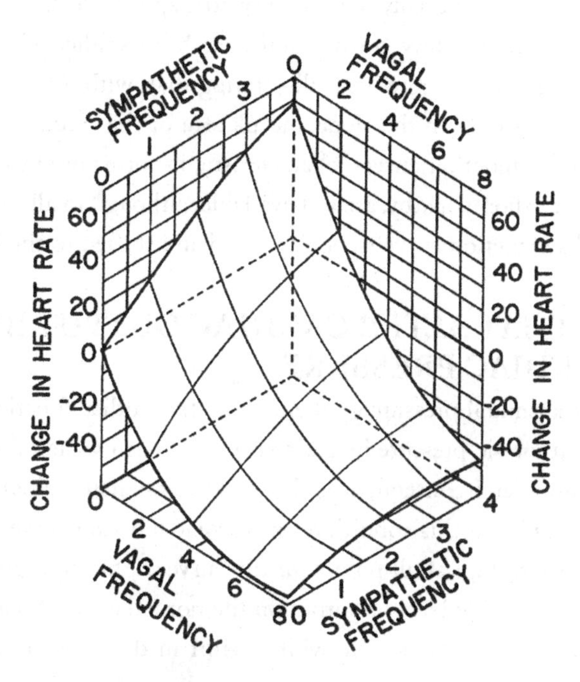

FIGURE 3.10: Three-dimensional interrelationships between sympathetic and parasympathetic stimulation of the heart on cardiac function. Reproduced with permission of Elsevier from reference [25], derived from data from reference [26].

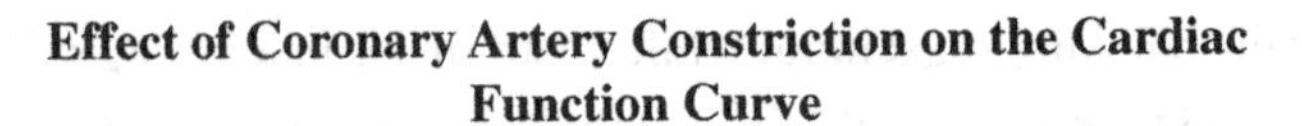

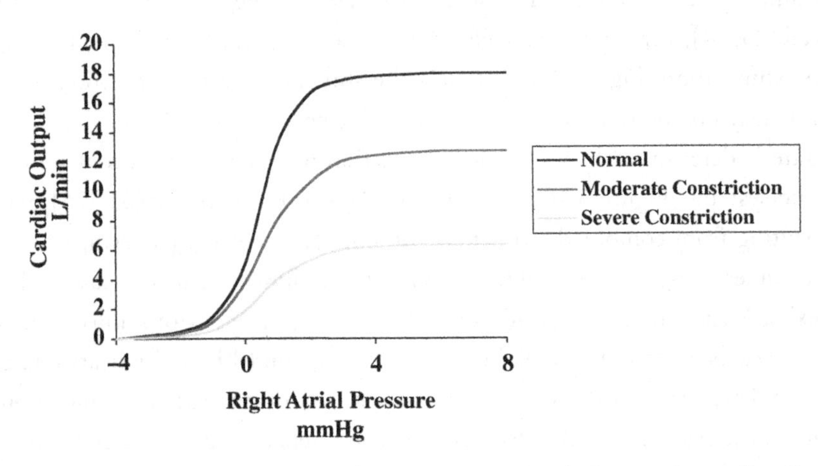

FIGURE 3.11: Impact of differing degrees of coronary artery occlusion on the cardiac function curve.

Myocardial disease or damage shifts the cardiac function curve to the right and downward. The effects of coronary vascular disease on cardiac function are complex, and it is difficult to make generalizations about the effects, especially so since very little applicable data exist. However, several experimental models of coronary artery constriction have been studied [27–30]. Curves derived from such studies are presented in Figure 3.11, illustrating that, with ischemic or hypoxic impairment of myocardial metabolism, both the slope and plateau of the cardiac function curve are depressed. Data describing the function curves' characteristic in forms of chronic heart failure, such as hypertrophy or dilated cardiomyopathy, are not available, although in all forms of cardiac failure, one may expect the cardiac function curve to be depressed and shifted to the left.

3.3 FACTORS THAT ALTER CARDIAC OUTPUT BY CHANGING EXTRACARDIAC PRESSURE

Reduction in the atrial transmural pressure gradient shifts the cardiac function curve to the left. If the chest is opened, intrathorasic pressure increases from its mean value of approximately −4 mm Hg to atmospheric pressure, and consequently, the pressure gradient across the wall of the right atrium decreases by 4 mm Hg. In this condition, the cardiac function curve is shifted in a parallel manner to the right by 4 mm Hg; at all points on the curve, right atrial pressure must be 4 mm Hg higher to achieve a comparable level of output on the normal curve. Breathing against positive pressure shifts the curve to the right as well, while a shift in the opposite direction results from breathing again negative pressure. The plateau value of cardiac output is not affected by changes in the pressure gradient across the atrial wall.

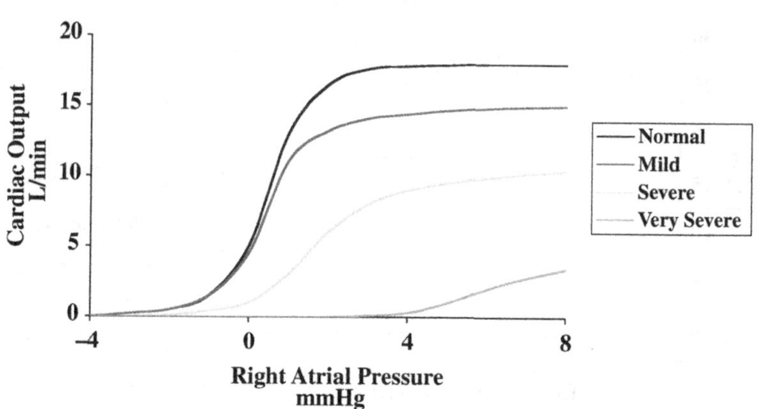

FIGURE 3.12: Effects of cardiac tamponade on the cardiac function curve.

Cardiac tamponade reduces the slope of the cardiac function curve. Cardiac tamponade is a condition caused by accumulation of fluid in the pericardium due to infection, trauma, or hemorrhage. Because the pericardium has low compliance, the increase in volume caused by the fluid increases intrapericardial pressure and reduces the atrial transmural pressure gradient. As a result, the cardiac function curve is shifted to the right, indicating that a higher level of atrial pressure must be reached in order attain a given level of cardiac output on the normal function curve. The situation is somewhat similar to the effect of increased intrapleural pressure or positive pressure breathing, except that in the case of cardiac tamponade, the pressure outside the atrium is not constant; rather, due to the low pericardial compliance, the pericardial pressure increases as cardiac volume and pericardial fluid volume increase. Cardiac volume increases as atrial pressure increases; therefore, as atrial pressure rises, pericardial pressure increases, limiting the rise in atrial transmural pressure gradient. Cardiac function curves produced in various degrees of tamponade are presented in Figure 3.12. The severity of tamponade is related to the volume of fluid and the pressure within the pericardium. Notice that the slope of the curves is reduced progressively as atrial pressure increased due to the progressive expansion of the heart. Furthermore, the plateau is reduced and may not be achieved at all at physiological levels of atrial pressure.

3.4 SUMMARY

The pumping of the heart working as an integral part of the cardiovascular system can be described quantitatively by function curves relating cardiac output as the dependent variable to right atrial

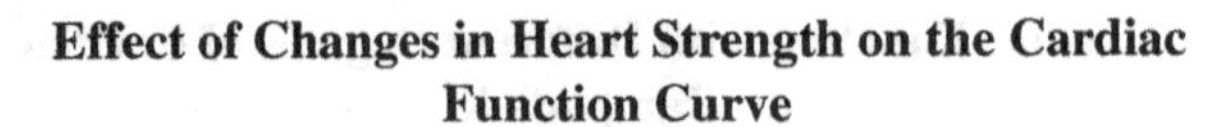

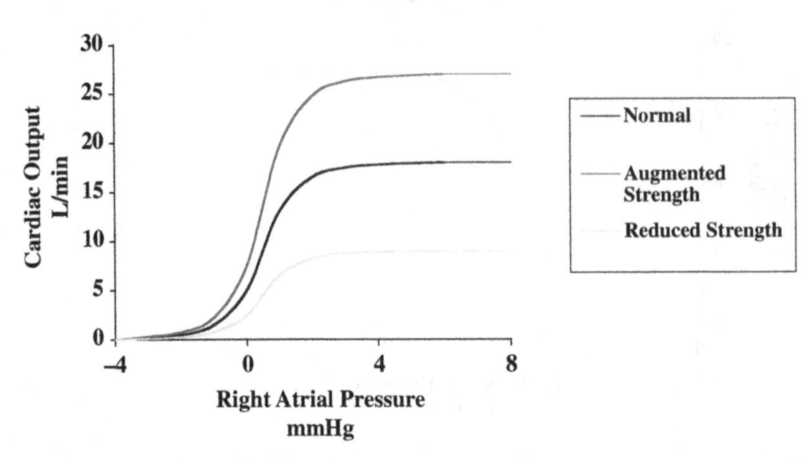

FIGURE 3.13: Effects of intrinsic changes in pumping ability of the heart on the cardiac function curve.

Effect of Extracardiac Pressure on the Cardiac Function Curve

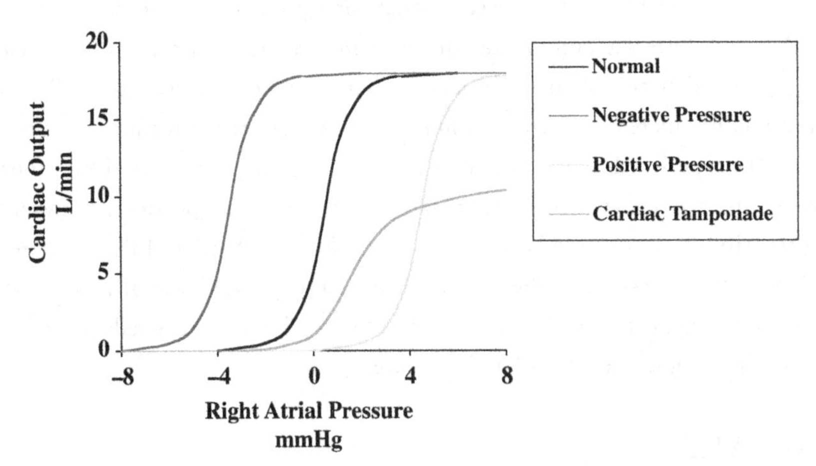

FIGURE 3.14: Effects of pressure outside the heart on the cardiac function curve are illustrated here.

pressure as the independent variable. Regardless of its condition or the factors affecting it, the heart's function in the system is accurately reflected by its cardiac function curve. Furthermore, cardiac output can be predicted from right atrial pressure if the cardiac function curve is known.

Cardiac function curves can be altered by two types of factors: changes in the effectiveness of the pumping ability of the heart and changes in extracardiac pressure. The family of cardiac function curves resulting from the effects of changes in strength and effectiveness of pumping is shown in Figure 3.13, and the family of curves resulting from changes in extracardiac pressure is presented in Figure 3.14.

• • • •

CHAPTER 4

Integrated Analysis of Cardiac Output Control

4.1 GRAPHICAL ANALYSIS OF CARDIAC OUTPUT REGULATION BASED ON COMBINED VENOUS RETURN AND CARDIAC FUNCTION CURVES

Regulation of cardiac output requires interaction of factors affecting the return of blood to the heart from the peripheral tissues and factors affecting the pumping ability of the heart itself. The previous two chapters described ways to develop quantitative descriptions of venous return and cardiac output as functions of right atrial pressure. Both function curves are graphical expressions of complex algebraic equations, each having the same two variables, flow and right atrial pressure. While neither alone can be solved for either flow or right atrial pressure, solving the two expressions by plotting both simultaneously on the same axes can yield solutions for both. Plotting the two relationships on the same right atrial pressure versus flow coordinates reveals that they intersect at the equilibrium point that is the cardiac output and right atrial pressure for the cardiovascular system described by those venous return and cardiac output functions. Figure 4.1 presents the combination of a venous return curve and cardiac function curve plotted together, intersecting at the equilibrium point [31].

Essentially, all factors that may affect the regulation of cardiac output alter the venous return and cardiac output relationships in the limited number of ways that were described in the previous chapters. Strengthening or weakening the heart's pumping ability can only increase or decrease the slope or plateau of the cardiac function curve, and all effectors of venous return can only alter the plateau, the slope, or the intercept of the venous return curve. Therefore, each action or stimulus affecting the cardiovascular system can be understood in terms of its effects on the two function curves, and if the effects are known, the resulting cardiac output can be determined.

Cardiac output responses to changes in mean systemic pressure and resistance to venous return can be determined graphically. Increasing or decreasing mean systemic pressure will shift the venous return curve to the left or right without affecting its slope. For example, if the cardiovascular system described by the function curves shown in Figure 4.1 received an input such as a rapid

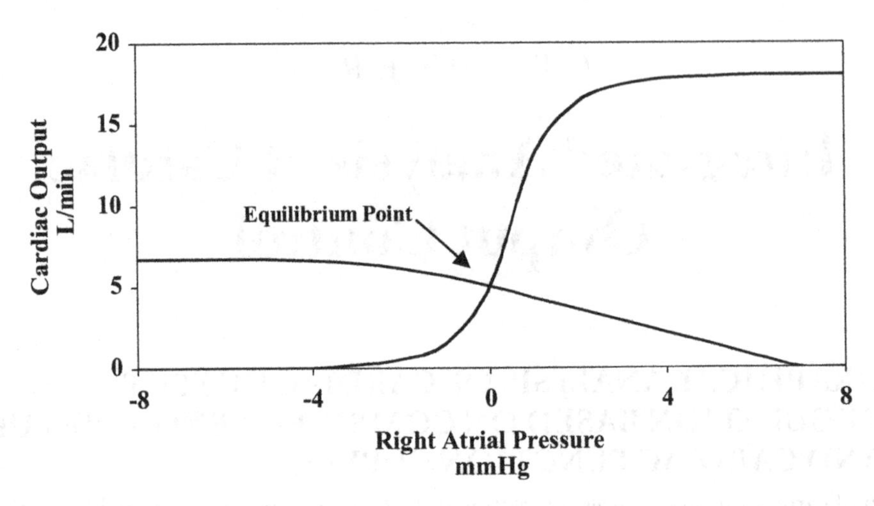

FIGURE 4.1: Presented in the figure are a normal venous return curve intersecting a normal cardiac function curve at the equilibrium point.

transfusion of blood that raised mean systemic pressure from 7 to 10 mm Hg, the venous return curve would be shifted parallel to the right, as shown in Figure 4.2 as the blue line. Consequently, the new venous return curve would intersect the normal cardiac function curve at a new equilibrium point, approximately 7.5 L/min and 0.7 mm Hg right atrial pressure.

If resistance to venous return was increased by 20%, as a result of reduction in metabolic demand by the body, the slope of the venous return curve would be reduced but the *x*-axis intercept would remain unchanged, as shown in the yellow line in Figure 4.2. The solution for the values of cardiac output and right atrial pressure of the cardiovascular system under these conditions would be the intersection of the normal cardiac output curve and the new venous return curve, where cardiac output is 4.4 L/min and right atrial pressure is –0.5 mm Hg.

In exercise, increased metabolic demand of the body may decrease resistance to venous return, and elevated sympathetic nervous system activity may raise the mean systemic pressure. If resistance to venous return were decreased by 50% and mean systemic pressure were increased to 12 mm Hg, the slope of the venous return curve would be increased, and the *x*-axis intercept would be shifted 5 mm Hg to the right, yielding the venous return curve illustrated in Figure 4.2 as green line. The cardiac output and right atrial pressure under these conditions would be the intersection of the new venous return curve and the normal cardiac output function curve, at which cardiac output is 14.4 L/min and right atrial pressure is 2 mm Hg.

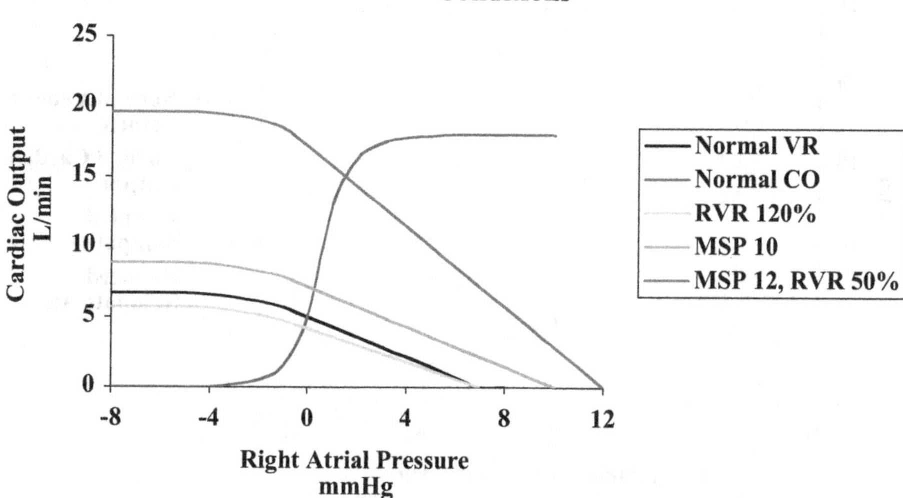

FIGURE 4.2: A normal cardiac function curve intersecting the normal venous return curve and venous return curves obtained following events that altered resistance to venous return and/or mean systemic pressure.

Cardiac output responses to changes in the pumping ability of the heart can be determined graphically. Nearly all effectors of cardiac pumping ability shift the function curve and/or the plateau. Raising sympathetic nervous system activity, for example, increases the slope and raises the plateau, while sympathetic blockade and parasympathetic stimulation have the opposite effect on both properties of the function curve.

If emotional excitement increased sympathetic stimulation by 50%, the new cardiac function curve would be shifted to the left, and the plateau would be increased compared to the normal curve. In Figure 4.3, the normal cardiac function curve and the normal venous return curve are plotted as dark blue lines intersecting at 5.0 L/min cardiac output and 0 mm Hg right atrial pressure, and the new cardiac curve is plotted as the yellow line, which intersects the normal venous return curve at a point where cardiac output is only slightly greater than normal and right atrial pressure is slightly below normal.

Conversely, if a pharmacological antagonist of the sympathetic nervous system transmitters were administered reducing the sympathetic nervous system effect on the heart by 50%, the pumping ability of the heart would be limited, yielding the function curve shown as the red line in Figure 4.3. It intersects the normal venous return curve where cardiac output is only slightly reduced from normal, and right atrial pressure is slightly greater than 0 mm Hg.

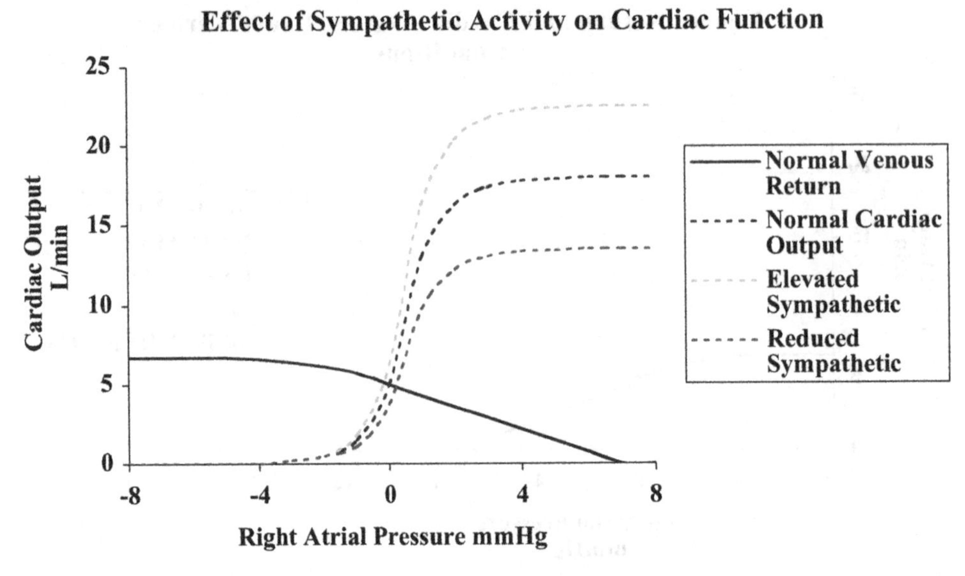

FIGURE 4.3: Normal venous return curve intersecting the normal cardiac function curve and cardiac function curves obtained after increased or decreased sympathetic stimulation of the heart.

Cardiac output changes can be determined graphically when multiple factors affect simultaneously both venous return and cardiac pumping ability. Most challenges faced by the cardiovascular system entail simultaneous actions on venous return and cardiac pumping ability. The effect of such complex changes can also be solved using graphical methods.

If a motor vehicle accident results in lacerations causing great pain and hemorrhage of several hundred milliliters, mean systemic pressure may fall to 5 mm Hg, resistance to venous return may increase 33% greater than normal, and cardiac pumping ability may increase to 125% of normal. What would be the new levels of cardiac output and right atrial pressure in this circumstance? The answer is not intuitively obvious, but it can be estimated by analyzing the effects of the perturbations on the venous return and cardiac function curves and plotting the functions together graphically to obtain the new equilibrium point. In Figure 4.4, the normal function curves are plotted as dark blue lines (venous return as solid lines and cardiac output as dashed), and the new relationships are presented as red lines. The venous return curve can be plotted knowing the mean systemic pressure, which is the *x*-axis intercept, 5 mm Hg, and the slope (1/resistance to venous return = 1/1.33), which is decreased to 75% of normal, approximately 0.5 L/min/mm Hg. The cardiac pumping ability, which is 125% of normal, may be represented as a function curve having a plateau value 22% higher than normal (22 L/min) but achieved at the normal plateau value of right atrial pressure, approximately 4 mm Hg. The intersection of the new curves occurs at a point where cardiac out is approximately 2.8 L/min and right atrial pressure is −0.7 mm Hg.

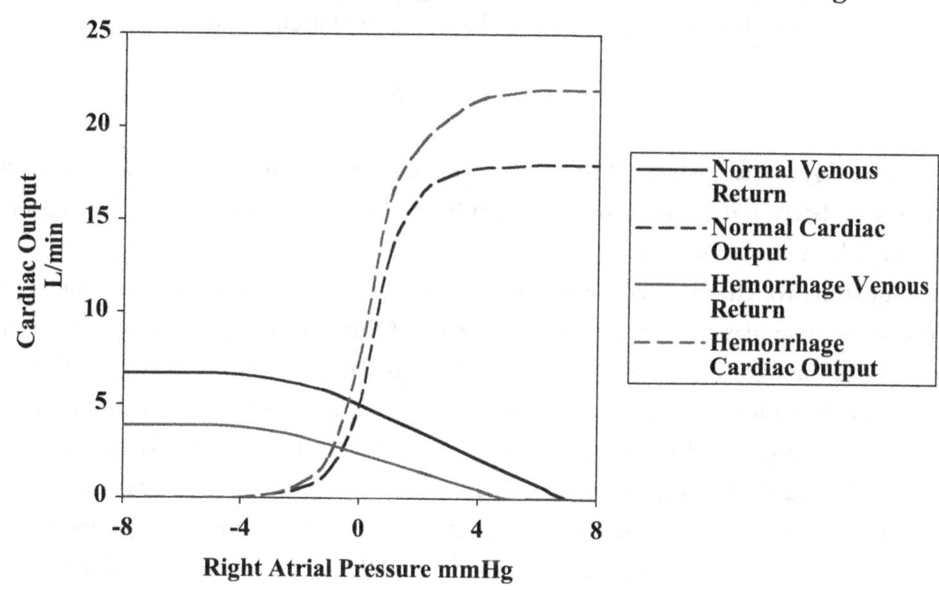

FIGURE 4.4: Cardiac output and venous return curves under normal conditions and following trauma and hemorrhage.

4.2 ALGEBRAIC ANALYSIS OF CARDIAC OUTPUT REGULATION

Investigators attempted in the past 100 years to derive algebraic expressions useful in analyzing cardiac output regulation. Some are more useful than others.

Cardiac output can be expressed as a function of heart rate and stroke volume. This is intuitively obvious and is frequently written as an equation:

$$\text{CO} = (\text{heart rate})(\text{stroke volume}) \tag{4.1}$$

The equation is useful in some circumstances in predicting cardiac output when only heart rate changes. However, most responses of the cardiovascular system involve changes in variables not included in the equation, especially those that determine stroke volume. Furthermore, the equation has led to some misunderstanding of regulation of cardiac output.

Ohm's law can be adapted to expressions of cardiac output regulation. The most straightforward adaptation is the following:

$$\text{CO} = P_a/R_s \tag{4.2}$$

where P_a is the systemic arterial pressure and R_s is the systemic resistance. This equation assumes that right atrial pressure is 0. The expression can be made more useful by including it as a variable:

$$CO = (P_a - P_{ra})/R_s \tag{4.3}$$

where P_{ra} is the right atrial pressure. The formula can be rearranged to find systemic resistance, but its usefulness in determining cardiac output is limited by simultaneous changes occurring in more than one variable in most situations.

The quantitative effects of changes in venous resistance on venous return are much greater than proportionally similar changes in arterial resistance. Ohm's law can be useful in understanding the flow of fluid through rigid tubes, but because the circulatory system is made up of compliant vessels, Guyton found it was necessary to modify Ohm's law in order for it to be useful in analysis of the cardiovascular system. Instead of using one simple term for systemic resistance, arterial and venous resistances had to be weighted separately according to the capacitances of the two segments of the vascular system [32]. The need for the modification is due to the arrangement of arterial resistance, R_a, and venous resistance, R_v, in series and each being positioned following capacitance segments, whose volumes are functions of the arterial pressure and capacitance and venous pressure and capacitance, respectively. Extra volume in each segment, EV_a and EV_v, is the contained volume greater than the unstressed volume. The series arrangement of the arterial and venous resistances and capacitance sections is presented schematically in Figure 4.5.

The derivation of the expression incorporating the capacitance-weighted arterial and venous resistances begins by restating arterial and venous pressures separately. The pressure in the arteries, P_a, is equal to right atrial pressure, P_{ra}, plus the pressure drop from the root of the aorta to the right atrium; the pressure in the veins is equal to venous pressure, P_v, minus the right atrial pressure. Pressures in the arteries and veins are defined by the product of cardiac output and the separate arterial and venous resistances, R_a and R_v:

$$P_a = CO(R_v + R_a) + P_{ra} \tag{4.4}$$

$$P_v = CO(R_v) + P_{ra} \tag{4.5}$$

The extra volume, EV, in the arterial and venous segments is equal to the pressure in that segment times its capacitance, C_a or C_v:

$$EV_a = [CO(R_v + R_a) + P_{ra}]C_a \tag{4.6}$$

$$EV_v = [CO(R_v) + P_{ra}]C_v \tag{4.7}$$

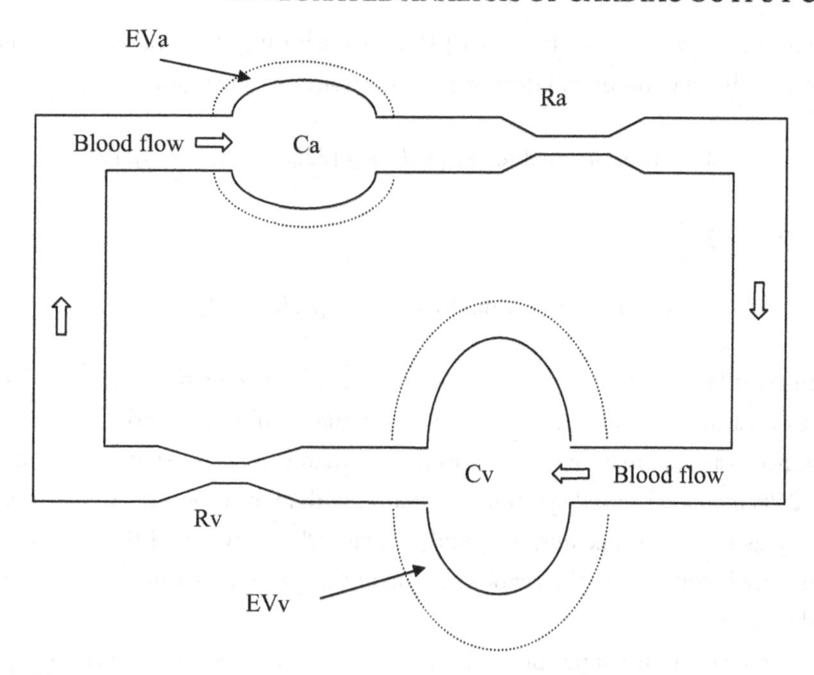

FIGURE 4.5: Vascular capacitance and resistance in the arterial and venous portions of the circulation arranged in series. The dashed lines around the capacitance segments represent volume greater than the unstressed vascular volume.

The mean systemic pressure is equal to the extra volume greater than the unstressed volume of the total systemic vascular system, EV_{syst}, divided by the systemic capacitance, C_s. The extra volume EV_s is equal to the sum of E_{va} and EV_v, and C_s is the sum of C_a and C_v:

$$P_{ms} = EV_s/C_s = (EV_v + E_{va})/(C_v + C_a) \quad (4.8)$$

Substituting the expressions for EV_v and EV_a derived in Equations (4.6) and (4.7) into Equation (4.8) and rearranging to solve for cardiac output yields the following:

$$CO = (P_{ms} - P_{ra})/\{[R_v C_v + C_a(R_v + R_a)]/(C_v + C_a)\} \quad (4.9)$$

Equation (4.9) is very helpful in understanding complex cardiovascular phenomena, and many of these will be analyzed in subsequent chapters of this presentation. But in the context of analyzing factors affecting venous return, the denominator makes clear the potentially preponderant effect of venous resistance on venous return and cardiac output. The importance of venous resistance is even more striking when viewed in light of the experimental data, indicating that the ratio of venous to arterial capacitances may be as much as 18:1 [3]. By substituting these values into Equation (4.9), we can appreciate the magnitude of effects on venous resistance changes:

$$\text{CO is proportional to } (P_{ms} - P_{ra})/\{[18R_v + 1(R_v + R_a)]/19\} \quad (4.10)$$

Simplification yields:

$$\text{CO is proportional to } (P_{ms} - P_{ra})/[R_v + (R_a/19)] \quad (4.11)$$

Therefore, at least under some conditions, a change in venous resistance may have a 19-fold greater effect on cardiac output than the same percentage change in arterial resistance. Normally, arterial resistance is approximately seven times greater than venous resistance. Consider the consequences of a 20% increase in total systemic resistance, with all of it occurring in arterial resistance. Calculated venous return/cardiac output would decrease 6%. However, if the 20% increase in total systemic resistance is confined to the venous portion of the system, the calculated decrease in venous return would be 53%.

While the quantitative superiority of changes in venous resistance over changes in arterial resistance in affecting cardiac output and venous return is impressive, it should be considered in the context of the usual operation of the cardiovascular control system. In normal conditions of health, arterial resistance is much more dynamic than venous resistance; arterial resistance is strongly affected by the sympathetic nervous system, circulating vasoactive hormones, such as epinephrine, angiotensin II, and PGE_2, as well as by local tissue autoregulatory mediators, especially those sensitive to reduction in tissue pO_2. Venous resistance, on the other hand, is comparatively less sensitive to the effects of these factors. Therefore, changes in arterial resistance frequently may contribute to the greater share of changes in total systemic resistance.

Cardiac output can be expressed as a function of capillary pressure, right atrial pressure, and venous resistance. Guyton proposed that cardiac output could also be expressed as a function of the pressure gradient from the mid-point of the capillaries to the right atrium divided by the vascular resistance from the capillary mid-point to the right atrium [32]:

$$\text{CO} = (P_c - P_{ra})/R_v \quad (4.12)$$

The terms of Starling's law of the capillary, which states that capillary pressure is equal to tissue pressure plus plasma colloid osmotic pressure minus tissue colloid osmotic pressure, can be substituted in Equation (4.12) to yield:

$$\text{CO} = (P_t - P_{cop} - P_{cot} - P_{ra})/R_v \quad (4.13)$$

in which P_t is the tissue fluid pressure, P_{cop} and P_{cot} are plasma and tissue colloid osmotic pressure, respectively, and P_{ra} is the right atrial pressure. This expression of cardiac output, like Starling's law of the capillary, is only valid in steady-state conditions; if one determinant of capillary pressure changes, several hours may be required for equilibrium to be reestablished.

The value of Equation (4.13) is derived from the independence of the determinants of capillary pressure from all other circulatory variables and from the other elements of the equation; colloid osmotic pressure in the plasma and extracellular space are functions of protein concentrations in the two compartments, tissue pressure is a function of the physical characteristics of the extracellular space and the volume of fluid it contains, right atrial pressure is primarily a function of the pumping ability of the heart, and resistance to venous return is determined by the physical characteristics of the venous vessels and the viscosity of the blood. This independence of the variables improves the probability that the equation is a valid and useful algebraic expression of cardiac output.

Equation (4.13) has several interesting and important implications. For example, it implies that cardiac output varies as a direct function of both tissue fluid pressure and plasma colloid osmotic pressure, with cardiac output increasing with increases in either variable. Additionally, the equation states that cardiac output varies inversely with right atrial pressure and resistance to venous return and that it is independent of arterial pressure except in conditions where afterload significantly affects right atrial pressure.

The equation can also be helpful in understanding regulation of cardiac output in heart failure. Frequently, patients with signs of severe congestive failure will have normal cardiac output at rest while having high right atrial pressure. These patients are also known to have high levels of tissue fluid pressure, which would contribute to elevated capillary pressure. Consequently, even with high right atrial pressure, the patients may have, according to Equation (4.13), a pressure gradient for venous sufficiently elevated to maintain normal cardiac output.

4.3 SUMMARY

Because regulation of cardiac output requires interaction of factors affecting return of blood to the heart from the peripheral tissues and factors affecting the pumping ability of the heart itself, both must be considered in attempts to analyze cardiac output control. Plotting venous return curves simultaneously with cardiac function curves is an effective means of considering the numerous important variables' roles in regulation of output. Most algebraic expressions of cardiac output deal explicitly with only factors affecting venous return, while the pumping function of the heart contributes indirectly via its effect of right atrial pressure. The most significant equations are based on Ohm's law, in which cardiac output is equated to the pressure gradient for venous return divided by the resistance to venous return. Differences between equations result from differences in expressing the pressure gradient and the resistance.

• • • •

CHAPTER 5

Analysis of Cardiac Output Regulation by Computer Simulation

5.1 RATIONALE FOR BUILDING AND USING MATHEMATICAL MODELS OF THE CARDIOVASCULAR SYSTEM

Models are developed beginning with hypotheses concerning the organization of the system and the function of its components. For example, to develop a cardiovascular model useful for analyzing hypotheses concerning the importance of factors affecting venous return in cardiac output regulation, the model could be organized with emphasis on components related to venous return. The components' functions could be described mathematically from data collected experimentally or from clinical observations. The function curves described in earlier chapters that were derived from experimental data could be expressed by one or more simple algebraic equations solvable by a digital computer. The components' equations could then be arranged in series with the solution of the first, the dependent variable, being the independent variable in the next, and the solution to the final equation being the independent variable in the first equation, forming a loop that could be solved iteratively by the computer. Each iteration may be considered a unit of time chosen by the investigator. The inclusion of time as a component of the simulation gives this technique an advantage over other methods in that the function of the system can be studied at more than one point in time after a change in conditions.

To begin a simulation, variables are assigned initial values, often normal values. The simulation can then be started with the series of equations solved iteratively with the value of selected variables listed numerically or plotted graphically versus time. The investigator may choose to have the variables' values expressed after each iteration or at longer intervals. After an initial period of operation, the investigator may introduce a change in the system designed to represent an experimental manipulation or a pathological condition. For example, to simulate a hemorrhage, a function could be introduced into the series of equations that would subtract a specified amount of blood in each iteration, representing a blood loss at a rate such as 10 mL/min. The responses of the variables of the simulated system over the subsequent period (number of iterations) could then be observed.

The construction of a model requires the investigator to consider his hypotheses in careful detail, eliminating "fuzzy thinking" or "hand-waving" logic. To write down the steps in the logic as a series of equations, the investigator must be able to think clearly through the details of a proposed hypothesis, frequently with greater rigor than when considering an idea mentally or when writing in prose.

Simulations enable the investigator to test the validity of hypotheses. A finding of disagreement of simulation solutions with actual system responses suggests that the hypotheses inherent in the model may not be valid. Often, such a situation is highly instructive, and by carefully studying the source of the model's error, the investigator may gain insight into the function of the system.

If the results of the model simulation agree with known responses of the actual system under consideration, the agreement supports, but does not prove, the validity of the hypotheses inherent in the model. To return to the hemorrhage example, if the simulated responses of significant variables, such as mean systemic pressure, right atrial pressure, arterial pressure, and cardiac output, to a hemorrhage of 10 mL/min agreed with known responses of the actual cardiovascular system to a similar rate of blood loss, the agreement could be viewed as favoring the tenets of the hypotheses. Agreement of many simulated responses of the model to responses of the actual system provides even greater support for the hypotheses' validity.

Model simulations permit careful observation of the components' responses with time. Once the soundness of the model has been amply supported and the investigator has reasonable confidence in the strength of the hypotheses, simulations can be used to analyze in great detail the temporal responses of any and all system components to challenges of the investigator's choice. Furthermore, the investigator can study simulations of events or situations that may be impossible to study experimentally. These processes provide an opportunity for heightened understanding of the workings of complex systems.

The inclusion of time as a variable in model simulations forces the investigator to recognize the significance of temporal responses of the system to challenges. Throughout the history of the study of cardiovascular physiology, nearly all experimental investigations were conducted over the course of a few hours. The data collected, usually at the conclusion of the study, were used to construct hypotheses assumed to be applicable to the long-term or steady-state operation of the cardiovascular system. But when constructing a mathematical model of the system, the investigator must consider how all variables change with time, and frequently, one may realize that cardiovascular responses require far longer to complete than was assumed previously. This realization led Guyton's laboratory to undertake studies of the cardiovascular system that extended over many days and weeks, requiring grueling effort but yielding valuable results.

5.2 CARDIAC OUTPUT ANALYSIS USING A SIMPLIFIED CARDIOVASCULAR MODEL

Beginning more than 40 years ago, Guyton and coworkers recognized the potential importance of using mathematical models of the circulatory system to analyze the control of cardiovascular function. They began their work before digital computers were generally available, doing initial simulations using analog machines. A most basic model they published from work with an early digital computer was based on the hypotheses that the heart pumped blood at the rate it returned from the circulation, without an appreciable change in right atrial pressure [33], hypotheses closely related to the Frank–Starling law of the heart [16]. Included were functions expressing the relationships between cardiac output, arterial pressure, total peripheral resistance, blood volume, and extracellular fluid volume. The model comprised the eight functions illustrated as block diagrams in Figure 5.1, arranged in a loop.

The first function, Block 1, represents the relationship between arterial pressure and urinary excretion; in the simplified model, excretion is of extracellular fluid. As arterial pressure increases, urinary output increases, with small increases in pressure causing large increases in urinary excretion.

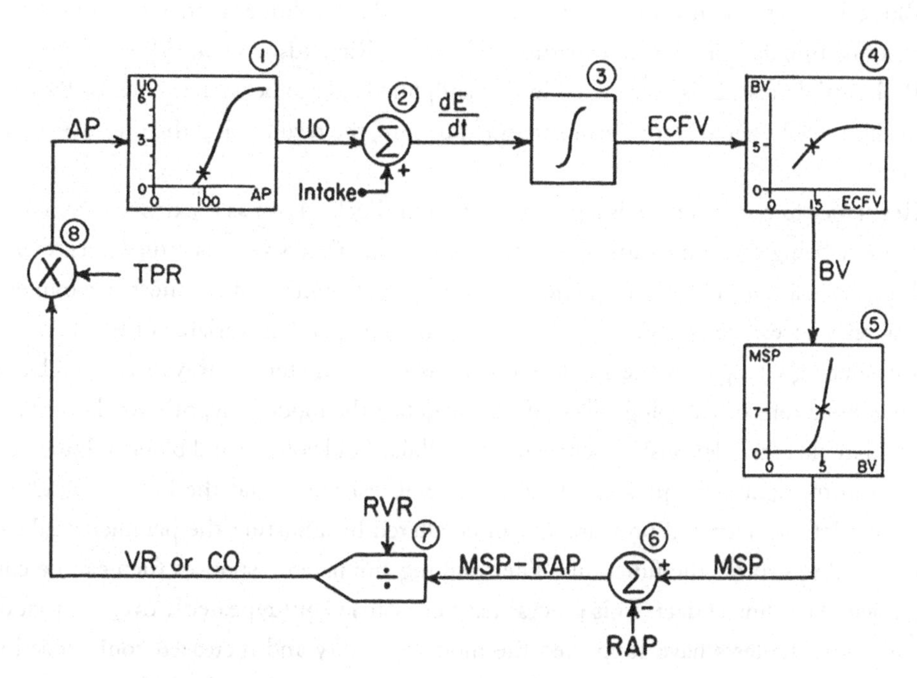

FIGURE 5.1: A block diagram of a mathematical model of a simplified representation of the cardiovascular system. Reproduced with permission of Elsevier from reference [34].

The quantization of the relationship was based on Selkurt's studies on kidneys perfused at controlled pressures [35]. The slope of the relationship is approximately a 6% increase in output for each 1 mm Hg increase in arterial pressure between 100 and 200 mm Hg.

In this simplified scheme, the rate of fluid intake is set as a parameter, a value that is fixed as a constant value that is not affected by the functions of the model, but can be changed by intervention of the investigator. The fluid intake rate is summated with the solution of the first function, the rate of urinary output, in the second function, Block 2. The output of this function is the rate of change of extracellular fluid volume. Initially, the rate of intake and excretion are equal to each other, and the subsequent variables in the feedback loop are unchanging, in a steady-state condition.

The rate of change of extracellular fluid volume is integrated over time in Block 3, the output being the extracellular fluid volume at a specific time point.

Block 4 expresses the relationship between changes in extracellular fluid volume and changes in blood volume. The function is complex, based on observations that increases in extracellular fluid volume produce increases in blood volume up to an extracellular fluid volume of approximately 22 L in a healthy man, but greater elevations in fluid volume are associated with edema formation with little further elevation of blood volume. The output of Block 4 is blood volume.

Block 5 depicts the function relating changes in blood volume to changes in mean systemic pressure, a function described from experimental data by Richardson et al. [8].

Right atrial pressure is considered in this simplified case to be a parameter. In Block 6, right atrial pressure is subtracted from mean systemic pressure, the result being the pressure gradient for venous return.

Resistance to venous return is a parameter that in Block 7 is set as the denominator, mean systemic pressure being the numerator. The output of this function is venous return or cardiac output.

Block 8 is a multiplication function, cardiac output multiplied by another parameter, total peripheral resistance. The result is arterial pressure, the independent variable of Block 1.

Introducing changes in the model's functions or parameters in ways that simulate experimental manipulations or pathological conditions can test the model's hypotheses. For example, the rate of fluid intake, the relationship between extracellular fluid volume and blood volume, resistance to venous return, right atrial pressure, total peripheral resistance, and the kidneys' fluid excretory function in relation to arterial pressure all can be altered by adjusting the parameter values or the equations used to express the functions. The resulting simulated responses then can be compared to experimental or clinical data. This process has been carried out repeatedly over the past decades, and in all cases, the tests have supported the model's validity and increased confidence in its accuracy. Consequently, this simple series of functions has been used as a basis for many, much more complicated cardiovascular models. It has also been used to gain understanding of basic principles of cardiovascular control.

The model describes a negative feedback control system that is fundamental to cardiovascular physiology. Although very few functions are included, along with several significant implications of complex physiological phenomena, the model illustrates key concepts in cardiovascular physiology. The importance of the pressure gradient for venous return as a determinant of cardiac output can be clearly appreciated in the model; changes in either mean systemic pressure or right atrial pressure resulting in reduction in the pressure gradient for venous return will reduce cardiac output. Reduction in blood volume from hemorrhage or dehydration reduces pressure gradient for venous return and consequently cardiac output in simulations and in experimental and clinical studies. Elevation of right atrial pressure also reduces pressure gradient for venous return, which is a prominent factor in the diminished cardiac output seen in cases of acute and chronic heart failure. The immediate effects of changes in total peripheral resistance on arterial pressure are to alter it in proportion to the change in total peripheral resistance, but the long-term simulated effects are more complicated, as discussed below.

Inclusion of renal excretory function in the model permits testing of the importance of changes in renal function on long-term circulatory control. The model predicts that if arterial pressure rises above the normal level, even by only a few mm Hg, for example, due to an infusion of a large amount of fluid, the rate of fluid excretion will rise to a level significantly greater than normal. If fluid intake remains at the normal level, the system will be in a state of negative fluid balance as long as arterial pressure is greater than normal, leading to a progressive reduction in extracellular fluid volume, blood volume, mean systemic pressure, pressure gradient for venous return, cardiac output, and finally arterial pressure. The effect will continue until arterial pressure falls back to the normal level, at which renal excretion will equal the rate of fluid intake and the system will again be in a stable condition with all variables at their normal values. Significantly, the simulations suggest that the correction of such a blood pressure rise would require several hours or more to complete, depending on the magnitude of the initial blood pressure rise. The prediction has been verified by experiments in which blood pressure has been raised or lowered by several different means, all confirming the operation of the negative feedback as proposed in the model.

The operation of this simple negative feedback loop has several implications that are important to understanding hypertension and blood pressure regulation in general. First, any factor that acutely increases blood pressure but does not affect the kidneys' ability to excrete fluid will not cause sustained hypertension. As long as renal function is normal with respect to fluid excretion, any increase in blood pressure will elicit operation of the negative feedback loop just described and, consequently, return blood pressure to the normal level. Therefore, although as discussed above, the immediate effect of an increase in total peripheral resistance on blood pressure is immediately apparent, the long-term response is more subtle if the vascular resistance change excludes renal resistance. Second, by analogous reasoning, sustained impairment in renal ability to excrete fluid will cause sustained hypertension. If the kidneys' function changes in a way that prevents fluid excretion at the normal rate at the normal blood pressure and fluid intake continues at a rate greater than excretion, a positive fluid balance will continue until arterial pressure rises to a level that increase renal fluid excretion to equal the rate of intake. Arterial pressure will remain at that level for as long as renal excretory function is impaired, requiring higher than normal perfusion pressure to balance excretion with intake of fluid. Similarly, augmenting renal ability to excrete fluid, which is the mechanism of action of diuretics and other antihypertensive medications, will reduce steady-state blood pressure.

5.3 CARDIAC OUTPUT ANALYSIS USING AN EXPANDED CARDIOVASCULAR MODEL

The model described above is very useful as a conceptual tool for studying basic cardiovascular function and the processes of modeling physiological processes. However, it is admittedly limited by its simplicity. Guyton and Coleman developed a more complex scheme built around the same fundamental core with additional components derived from the effects on the circulatory system of changes in heart strength, autonomic nervous system activity, and local vascular autoregulatory mechanisms. The block diagram of the more complex model is illustrated in Figure 5.2 [33].

The model contains 29 functions that reduce complex physiological relationships to simplified mathematical expressions. It illustrates the modeling process of incorporating additional complexity by incrementally adding functions around a central core. This is about the largest model that can be studied in block diagram form; illustrations of more complex models begin to resemble a maze. However, construction of this one can be followed graphically with some brief explanation.

Blocks 1–8 are the same functions used in the previous model. Its operation is affected by input from three negative feedback control loops: the local tissue autoregulatory system, Blocks 9–14; the effects of changes in heart strength on cardiovascular system, Blocks 16–19 and 27; and the effects of changes in autonomic nervous system function on circulatory control, Blocks 20–29.

5.3.1 Local Tissue Autoregulation

Previously, local tissue autoregulation was discussed with regard to its effect on vascular resistance (Chapter 2). Most tissues have the inherent capacity to regulate their flow of blood to meet metabolic needs. If flow is inadequate for their current requirements, local mechanisms within the resistance vessels of the tissue respond by causing vasodilation of the vessels regulating flow to the capillaries of the tissue, thereby increasing tissue blood flow and nutrient delivery. If flow is greater than required, the same control mechanisms act to increase vascular resistance and decrease flow. Block 9 depicts the function relating cardiac output as the independent variable to the rate of change in tis-

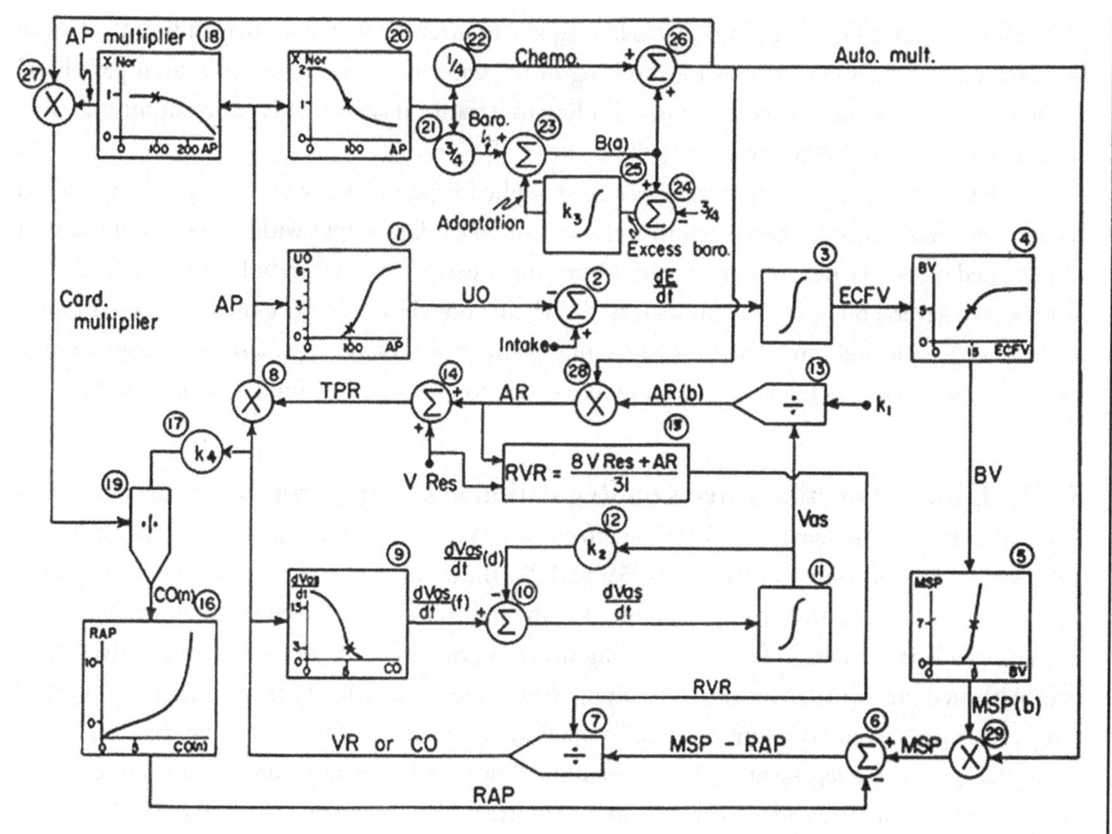

FIGURE 5.2: A block diagram of a mathematical model of a more complex representation of the cardiovascular system with additional components added around the basic eight-function loop. Reproduced with permission of Elsevier from reference [36].

sue vascularity, representing vascularity of all tissues. As cardiac output decreases, the reduced flow through the tissues of the body increases the value of the vascularity variable. Blocks 10–12 are a modeling technique used to integrate the change in the vascularity variable and provide a time delay in the response determined by a time constant, parameter K2, in Block 12. The product of Block 11, vascularity, divides the value of parameter K1, representing the effect of metabolic rate on vascular resistance, in Block 13, yielding arterial resistance. In Block 14, arterial resistance is multiplied by a factor from the autonomic nervous system functions, the product of which is summated in Block 14 with the parameter representing venous resistance, V_{res}, yielding total peripheral resistance. The arterial resistance variable is also used along with V_{res} to determine resistance to venous return, in Block 15, which is an equation derived from the ratio of arterial to venous resistance and the relative contribution of each to resistance to venous return (Chapter 2). In the model, the effects of a value of whole body tissue flow rate (CO) that is inadequate to meet the metabolic demands of the tissues is an increase in the vascularity factor resulting in decreases in arterial resistance, total peripheral resistance, and resistance to venous return, leading to iterative increases in cardiac output that ultimately meet the blood flow demands of the tissues.

The local autoregulatory system has been studied experimentally and can produce marked reductions in resistance to venous return and increases in cardiac output within seconds of the start of increased metabolic demand in exercise. Over longer periods, the magnitude of the effect can be as great as a several hundred percent increase in whole body flow. The system's temporal responses are both very rapid and can be sustained indefinitely; the rapid response is caused by vasodilation of existing vessels, whereas the long-term response is due to growth of additional blood vessels.

5.3.2 Cardiac Function Effects on Regulation of Cardiac Output

A greatly simplified scheme to include the effects of changes in cardiac function on the cardiovascular system is represented in Blocks 16–19 and 27. Block 16 is a cardiac function curve plotted with venous return (or cardiac output) plotted as the independent variable and right atrial pressure as the dependent variable, so for this modeling situation, venous return determines right atrial pressure, with an increase in venous return resulting in an increase in right atrial pressure. In the normal range, increases in venous return cause only small increases in right atrial pressure. Block 17 is an adjustable parameter representing heart strength. Block 18 is the function relating the effect of arterial pressure, or afterload, on cardiac output; as arterial pressure increases to high levels, cardiac output is impaired. The effect is combined in Block 27 with the autonomic effect on the heart, the product of which divides the heart strength factor in Block 19. The product of Block 19 is used to modify the cardiac the function curve in Block 16.

5.3.3 Autonomic Nervous System Effects on Cardiac Output Regulation

Blocks 20–29 represent the functions of the autonomic nervous system that interact with circulatory control and cardiac output regulation. Block 20 represents the relationship between arterial pressure as the independent variable and the autonomic factor as the dependent variable. Blocks 21 and 22 apportion the factor, with one-fourth going to the chemoreceptors and three-fourths to the baroreceptors. Baroreceptor drive to the autonomic system is known to adapt over time. A change in arterial pressure from the initial level causes a change in baroreceptor drive, but this adapts within a few days back to the level associated with the initial pressure. Blocks 23–25 model this adaptation phenomenon. Chemoreceptor drive does not adapt, and it is summated with the adapted baroreceptor output in Block 26, yielding the autonomic multiplier. The autonomic multiplier interacts with the other components of the model system in three areas: in Block 27, it multiplies the effect of arterial pressure on cardiac strength (Block 18); in Block 28, it multiplies the effect of the vascularity factor on arterial resistance; and in Block 29, the autonomic factor multiplies the effect of blood volume on mean systemic pressure (Block 5).

This more complex model, while still an extremely simplistic representation of cardiovascular function, enables testing and exploration of many more hypotheses than the first eight-function loop. Comparisons of experimental data and clinical observations with simulations in many varied circumstances have verified the model's validity for several decades.

Very significantly, this model, which is still at a basic level of complexity, clearly predicts the independence of long-term arterial blood pressure control from all cardiovascular functions except the relationship between renal perfusion pressure and renal excretory function. Manipulation of functions such as resistance to venous return, peripheral vascular resistance (excluding renal vascular resistance), vascular capacitance (compliance), heart strength (within normal limits that preclude heart failure), and autonomic nervous system function (excluding effects on renal function) will not affect the predicted long-term regulation of arterial pressure. These predictions have been tested in experimental studies in many animal models and in clinical studies for nearly half a century; data from each have confirmed the prediction and clearly demonstrated the dominance of the renal perfusion pressure–renal sodium and water excretion rate–body fluid volume negative feedback system in long-term blood pressure control. Alterations in other portions of the cardiovascular system have significant short-term effects on blood pressure, but those are overridden by the long-term ability of the renal body fluid volume regulatory mechanism to maintain arterial pressure at its control level.

5.4 DIGITAL HUMAN

Modeling of physiological systems has developed for several decades since Guyton and Coleman made their first cardiovascular efforts. Coleman's work has progressed to his current model, *Digital Human*, which is much more complex than earlier contributions, and includes many more physiological systems and greater detail. However, its cardiovascular functions closely resemble the core of the first models, with added units representing peripheral feedback control systems interacting with the central functions. Because the same fundamental cardiovascular functions persist in the current model, its simulations can be used to illustrate and test basic cardiovascular concepts and regulation of cardiac output.

Coleman designed *Digital Human* to be an open source publication, easily used by those interested in cardiovascular and broadly related areas of physiology. It can be downloaded from http://digitalhuman.org and used on office computers using Windows operating systems (a Macintosh version may be available in the near future). The reader is encouraged to make use of this valuable resource.

5.5 SUMMARY

Several aspects of modeling the cardiovascular system can be helpful in gaining an understanding of cardiac output regulation and circulatory control in general. The process of building the model requires the investigator to carefully think through the proposed hypotheses in sufficient detail to assemble a series of equations, each representing a component of the proposed system. Generally, it is very difficult to assemble mentally the details of complex proposals, and nearly impossible when several parallel and in-series components with differing time constants interact. Often, just the attempt to write down the details of hypotheses in preparation for the mathematical expressions is sufficient to disprove untenable ideas.

Having a solvable mathematical model representing a hypothetical system provides a means to test the proposed system by comparing the results of model simulations with experimental data and clinical observations. In many instances, such comparisons are the only way to test complex proposals. One case of nonconformity of simulated results with data from actual experiments is sufficient to reject a hypothesis and force the investigator to seek the errors of his thinking, "back to the drawing boards." Probably, most useful models are completed atop a heap of corrected errors. Model simulations can never definitively prove the validity of a hypothesis, but if model simulations are found to agree with data from a wide variety of experiments and clinical situations, investigators and other students of the subject can have some measures of confidence in the proposed hypothetical system. With each additional finding of agreement, the confidence increases.

Once a model's validity is firmly supported and the investigator has confidence that the model is an accurate representation of the actual system, its simulations can be studied to investigate the detailed functions and relationships of the cardiovascular system in ways that may not be possible in experimental physiological or clinical analyses. The investigator and students of the subject both can share this aspect of mathematical modeling.

• • • •

CHAPTER 6

Analysis of Cardiac Output Control in Response to Challenges

6.1 REFLEXES INITIATED BY BARORECEPTORS AND OTHER FACTORS

Cardiovascular responses to all challenges include participation of either or both components of the autonomic system. The impact of changes in autonomic activity on venous return and cardiac function were described previously in Chapters 2 and 3.

Sympathetic nervous system activity can change mean systemic pressure over the range of 5–18 mm Hg or more. The effect is due to constriction of the venous capacitance vessels as well as contraction of the spleen. In addition, sympathetic stimulation causes increased vascular resistance in the splanchnic vessels, transferring blood from the gut and liver into other parts of the vascular system. Similarly, the cardiac effect of sympathetic stimulation significantly reduces cardiac volume, producing a transfer of blood from the heart into the peripheral circulation. With complete inhibition of all sympathetic activity, mean systemic pressure falls from the normal level of 7 mm Hg to approximately 5 mm Hg, whereas maximal sympathetic stimulation can increase it to up to 18 mm Hg or greater [4, 11].

Vascular resistance is also controlled in part by the level of sympathetic activity. The majority of the effect is at the precapillary resistance vessels, the arterioles, and small arteries, with much less effect on the veins. Consequently, although changes in sympathetic nervous system activity can have very great effects on precapillary and arterial resistance, maximal sympathetic nervous system stimulation can only increase resistance to venous return by approximately 20% (see Chapter 2 for full explanation).

Figure 6.1 illustrates the effects of sympathetic stimulation of the peripheral vasculature without affecting the heart [37]. The normal function curves are presented as black lines intersecting at the point where cardiac output is 5 L/min and right atrial pressure is 0 mm Hg. Totally inhibiting sympathetic activity shifts the venous return curve downward and to the left (yellow curve) from its normal position due primarily to a reduction in mean systemic pressure from 7 to 5 mm Hg. With no sympathetic activity, the venous return curve intersects the unaffected cardiac output

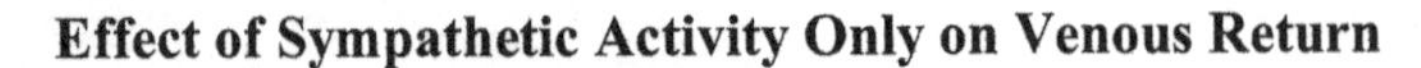

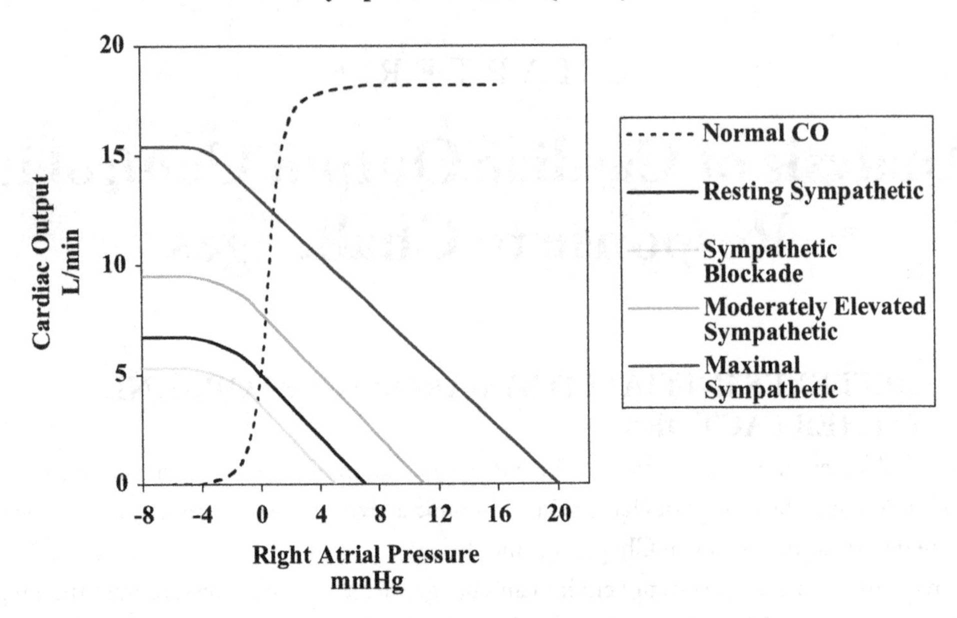

FIGURE 6.1: A normal cardiac output function curve and a series of venous return curves including the normal curve and others that would be observed under conditions of different levels of sympathetic stimulation of only the vasculature.

curve at the point where cardiac output/venous return is of 3.4 L/min, 32% less than the normal cardiac output of 5.0 L/min. Moderate sympathetic stimulation shifts the venous return curve to the right and upward (blue curve) as a result of the increase in mean systemic pressure to approximately 11 mm Hg. The equilibrium point at which venous return and cardiac output are equal is increased from the normal point to one with coordinates of 7.8 L/min and 0.5 mm Hg for cardiac output and right atrial pressure. Maximal stimulation (purple curve) shifts mean systemic pressure to 20 mm Hg and shifts the venous return curve upward, although the slope is reduced due to an elevation of resistance to venous return. Consequently, the venous return and cardiac function curves intersect at the point where cardiac output is 13 L/min. These data demonstrate that the range of sympathetic stimulation of the peripheral vasculature alone, without any cardiac effect, is capable of shifting cardiac output from two-thirds to more than double the normal level.

Under normal, resting conditions, sympathetic stimulation of the heart alone results in only small increases in cardiac output. Because the healthy heart has the capacity to increase its strength of contraction and pump as much blood as it receives in response to very small increases in right atrial pressure, increasing the heart's strength by sympathetic stimulation has little effect on output.

Sympathetic stimulation does shift the cardiac function curve to the left, but the normal curve has such a great slope near the normal operating point that additional strengthening can only produce a small increase in slope.

The magnitude of the sympathetic effect of solely cardiac stimulation is somewhat variable and dependent on several other factors. If parasympathetic activity is elevated, the cardiac function curve may be shifted to the right and the slope reduced; under these conditions, sympathetic stimulation may significantly affect the function curve's slope and plateau and strongly increase cardiac output. During very high levels of demand for cardiac output, for example, during very strenuous work or exercise, the level of cardiac output may be either on the plateau or in the upper range of the function curve where slope is reduced. Under these circumstances, the shift in the function curve resulting from sympathetic stimulation of the heart can greatly increase cardiac output.

The combined effects of sympathetic stimulation on venous return and cardiac function are illustrated in Figure 6.2. Three levels of stimulation are illustrated: normal (black), complete removal of sympathetic activity (blue), and maximal stimulation (red). The range of cardiac output/venous return resulting from zero to maximal stimulation predicted by this analysis is from approximately a 30% decrease to a 280% increase.

The pressoreceptor reflex can increase cardiac output by approximately 20%. The pressoreceptor reflex, also known as the arterial baroreceptor reflex, is elicited by reduction in stress in the walls of the carotid sinuses and aortic arch. Stretch-sensitive receptors in these structures initiate action potentials in nerve fibers that project to the brain stem. The rate of output from the receptors is proportional to the degree of stress, so that with reductions in blood pressure, firing rate from the receptors decreases. The autonomic nervous system responds to a reduction in impulse traffic from the receptors with an inversely proportional increase in sympathetic nervous system activity and a reduction in parasympathetic activity. Conversely, an increase in arterial pressure results in a reduction in sympathetic activity and an increase in parasympathetic fiber traffic. The effect of the reflex on short-term arterial pressure regulation is quite prominent and is especially important in man in maintaining arterial pressure in an upright posture. However, its effect on cardiac output regulation appears to be relatively minor. Carefully conducted animal experiments in numerous laboratories over many years yielded consistently variable results, with the greatest effect on cardiac output of maximal carotid baroreceptor reflex stimulation averaging approximately 10–15% [38–41]; however, within the experimental groups, it was not unusual to observe some animals whose cardiac output response was negative. The total reflex requires summation of input from both carotid and aortic receptors, with the aortic component contributing approximately one-quarter of the total. Allison [42] reported that reducing blood pressure in an isolated aortic arch from 100 to 0 mm Hg increased cardiac output by approximately 5%. Therefore, assuming that the inputs from the carotid

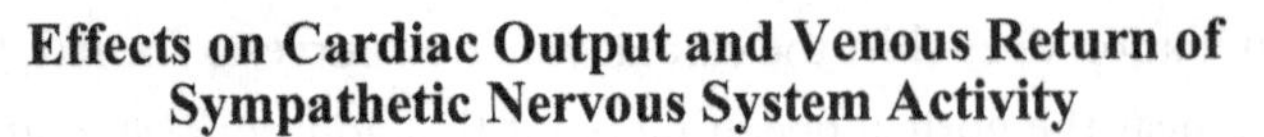

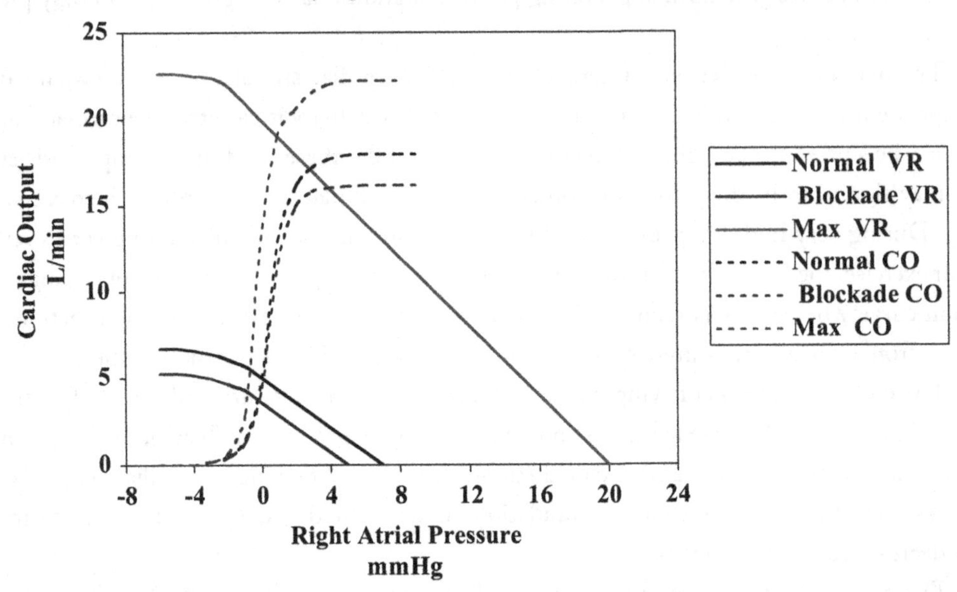

FIGURE 6.2: Expected results of different levels of sympathetic stimulation on both the venous return and cardiac function relationships.

and aortic receptors are simply summated by the autonomic system, the total maximal baroreceptor effect on cardiac output may be approximately a 20% increase.

The central nervous system ischemic reflex can elicit maximal sympathetic nervous system activation and a 100% increase in cardiac output. Ischemia in the brain, especially in the brain stem, results in hypoxia and hypercapnia and produces strong activation of sympathetic reflexes, a response referred to as the Cushing reflex. Asphyxia has the same effect.

These intense sympathetic reflexes have powerful effects throughout the cardiovascular system, raising mean systemic pressure to more than 20 mm Hg [11]. The cardiac output response to ischemia is not immediate but is preceded by a vagally mediated period of bradycardia lasting several minutes [43]. Once the hypoxia becomes severe, the cardiac output approximately doubles, even while mean arterial pressure rises to more than 200 mm Hg.

The vasculature of the brain has a highly effective autoregulatory mechanism that maintains blood flow at near the normal level over a wide range of arterial pressure, down to approximately 60 mm Hg. In severe hemorrhage, if arterial blood pressure falls to as low as 50 mm Hg, brain blood flow will fall to levels that elicit the central nervous system ischemic reflex, initiating maximal sympathetic reflexes throughout the cardiovascular system. Sagawa [21, 44, 45] estimated the reflexes resulting from hemorrhage to these levels to be four to six times more intense than those driven by maximal arterial baroreceptor activation.

Systemic hypoxia elicits central nervous system reflexes and peripheral vasodilation. A reduction in blood oxygen content or an impairment of cellular ability to utilize oxygen, as in metabolic poisoning, can have dramatic effects on the cardiovascular system. The autonomic nervous system response to hypoxia is similar to that described above for the Cushing reflex. As brain oxygen levels fall, sympathetic activation increases, providing cardiac strengthening and a tendency to increase peripheral resistance. At the same time, however, peripheral blood vessels dilate in response to the much stronger stimuli elicited locally as a consequence of local tissue hypoxia. If the degree of hypoxia is moderate, cardiovascular function could be described graphically, as shown by purple function curves in Figure 6.3 (function curves drawn in black lines represent normal venous return and cardiac functions). The cardiac function curve is shifted upward and to the left due to the sympathetic effect to increase the slope and the plateau. The venous return curve is shifted to the right as a result of the sympathetically mediated increase in mean systemic pressure and rotated clockwise due to the reduction in resistance to venous return associated with locally mediated vasodilation in

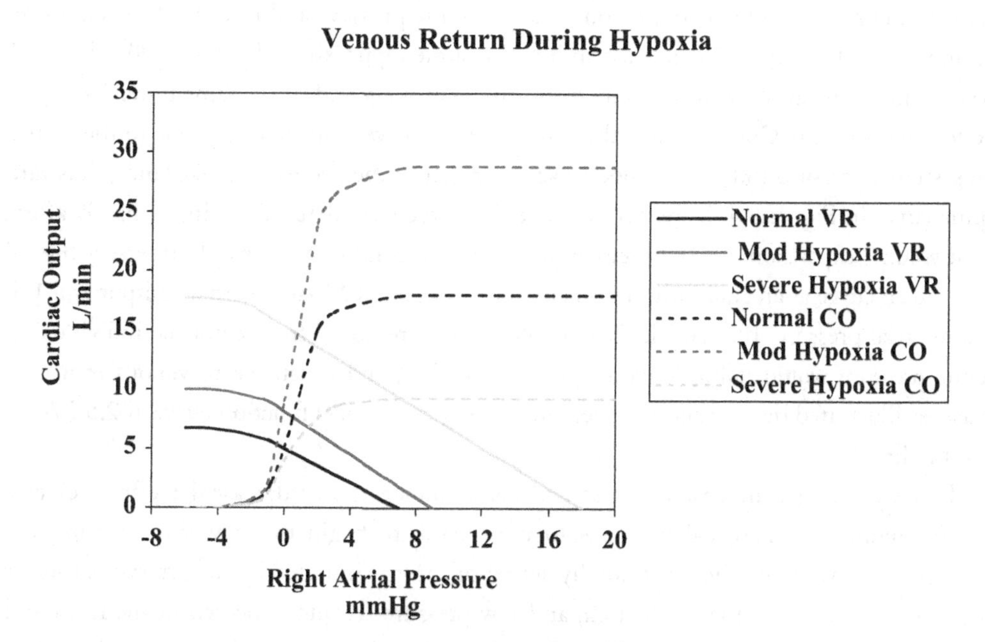

FIGURE 6.3: Effects expected of moderate and severe hypoxia on venous return and cardiac function curves. In severe hypoxia, cardiac function is impaired due to myocardial hypoxia, shifting the function curve downward and to the right.

response to inadequate oxygen delivery. The equilibrium point of intersection is at a cardiac output/venous return coordinates 8.5 L/min and 0 mm Hg.

Hypoxia has metabolic consequence that becomes increasingly severe as tissue oxygen levels fall. At some point, the negative cardiac consequences become more prominent than the positive inotropic effects of strong sympathetic stimulation, and the slope of the cardiac function curve begins to decline while the plateau level decreases. With severe hypoxia, cardiac function could be described by the yellow curve in the figure. But in the same condition, mean systemic pressure would be increased further by more intense sympathetic stimulation, and resistance to venous return would be maximally reduced due to intense tissue hypoxia. The venous return curve would be rotated clockwise and shifted to the right, intersecting the cardiac function curve at 9.3 L/min but at a greatly elevated right atrial pressure (7.0 mm Hg). Eventually, with progression of the level of hypoxia, left atrial pressure would rise to a value that would give rise to pulmonary edema, leading to death.

6.2 CHANGES IN BLOOD VOLUME

In Chapter 2, the relationships between blood volume and venous return were discussed, including the effect of changes in blood volume on mean systemic pressure and resistance to venous return. The immediate effect of a 14% increase in blood volume (approximately 900–1000 mL in a 70 kg man) is to double mean systemic pressure from 7 to 14 mm Hg and, at the same time, decrease resistance to venous return. Consequently, the venous return curve is shifted right due to the increase in mean systemic pressure and rotated clockwise as a result of the decrease in resistance, illustrated by the pink curve in Figure 6.4. If the cardiac function curve were determined immediately after such a blood volume expansion, it would remain in its normal condition but would intersect the shifted venous return curve at an equilibrium value of approximately 12 L/min cardiac output and 1.5 mm Hg right atrial pressure. If 5% of the blood volume were removed from a normal individual, mean systemic pressure would fall approximately 2–3 mm Hg, and resistance to venous return would increase, as illustrated by the yellow curve, intersecting the cardiac function curve at 2.5 L/min and −1.0 mm Hg.

Rapidly acting compensatory mechanisms attenuate the cardiovascular effects of changes in blood volume. Physiological compensatory mechanisms begin to act within seconds of large changes in blood volume. The most rapidly acting of these is initiated by the arterial baroreceptors in the carotid bodies and the aortic arch, and low pressure receptors located in the right and left atria. Elevations in pressure in the atria or arterial vessels due to blood volume expansion sensed by the receptors initiate reflex withdrawal of sympathetic nervous system activity as well as increases in parasympathetic activity, affecting the determinants of venous return and cardiac function. The

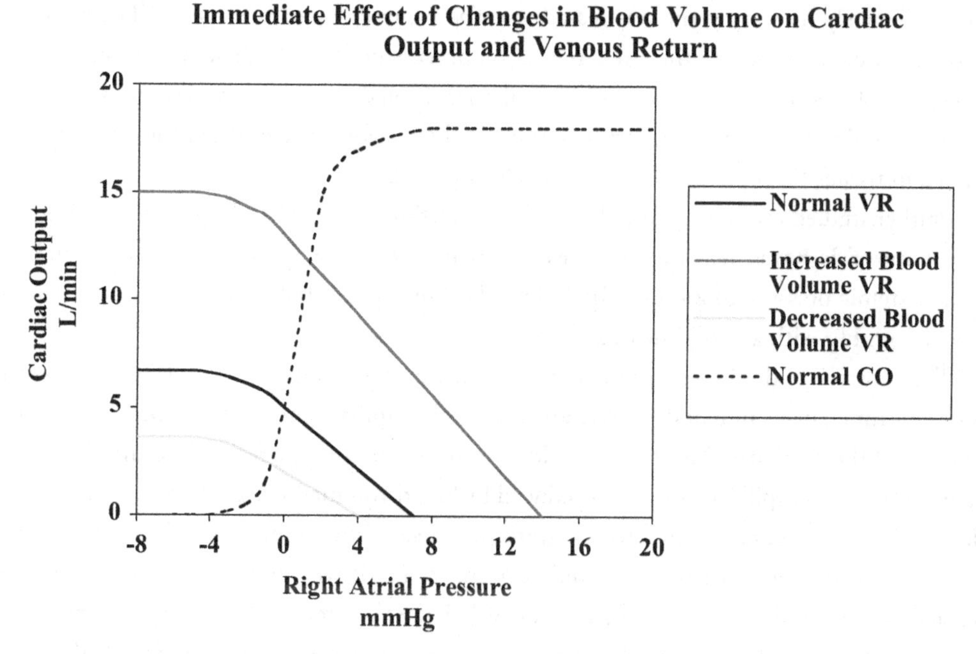

FIGURE 6.4: Expected effects of changes in blood volume on venous return and cardiac output.

reduction in sympathetic activity rapidly attenuates the rise in mean systemic pressure, preventing a portion of the rightward shift in the venous return curve. Simultaneously, the withdrawal of sympathetic tone and the elevation of parasympathetic activity reduce the slope of the cardiac function curve. Consequently, even before a large increase in blood volume may be achieved, effective compensatory mechanisms begin acting to limit the effects on the determinants of cardiac output/venous return. The immediate responses to sudden blood volume reductions would be mediated by the same systems and would be in opposite directions.

Stress relaxation of the vasculature begins within minutes of a change in blood volume, attenuating the effects of volume change on venous return. When blood vessels are stretched, wall tension immediately increases, but almost as rapidly, the tension begins to decline due to accommodation of active and passive elastic components of the vessel wall. When vessel stretch and wall tension suddenly decrease as a result of acute blood loss, the opposite processes take place, with wall tension increasing. These phenomena are referred to as *stress relaxation* and *stress relaxation recovery.*

When blood volume increases or decreases, wall tension of the vessels of the entire system is affected, either increased or decreased depending on the direction of the change in volume. In addition, tissues and organs that act as vascular reservoirs respond to blood volume changes, especially the liver and spleen. With increases in blood volume, they increase the volume of blood they contain, effectively increasing the unstressed vascular volume, V_0, of the system. The responses of these organs are due in part to the stress relaxation process per se and partially to sympathetic nervous system reflexes. The highly distensible capillaries throughout the body may also serve a similar function; if capillary pressure increases as a result of blood volume elevation, the volume of blood contained in tissues throughout the body may also increase.

Prather studied the time required for stress relaxation to occur [9]. In preliminary studies, the blood volume of dogs was increased by 35% by rapid infusion. The immediate effect was an increase in mean systemic pressure to 24 mm Hg, followed by an exponential decline to a value of approximately 10 mm Hg with a half-time of 2–4 min.

Transudation of fluid across capillary membranes buffers changes in blood volume. Associated with significant changes in blood volume are changes in capillary pressure that produce movement of fluid out of the capillaries into the interstium or in the opposite direction into the blood flowing through the vessel if capillary pressure is reduced by blood volume loss. This process begins as soon as blood volume changes and can have a significant effect within a few hours due to the extremely large capillary surface area across which fluid can move, and it will continue as long as the Starling forces across the capillary membranes are out of balance. Therefore, these mechanisms have the potential to return blood volume back to the initial value. However, fluid transudation alone usually does not have the opportunity to completely correct alterations in blood volume; fluid excretion by the kidneys is strongly enhanced by mechanisms activated by volume expansion, whereas volume loss is a strong stimulus for thirst, capable of restoring fluid volume in a very short time.

The combined effects of autonomic reflexes, stress relaxation, and fluid transudation in response to blood volume expansion are presented in Figure 6.5. The data are from Prather's work referred to above [9]. Responses to volume expansion achieved by rapid infusion of 500 mL of either Tyrode's solution, whole blood, or 5% dextran into anesthetized dogs were recorded for a 2-h period.

Notice that blood volume (panel A) in Tyrode's study returned to approximately the initial value by 80 min after volume expansion due to the rapid transudation of colloid-free fluid out of the vascular space. In the whole blood and dextran groups, blood volume remained elevated after 120 min, although 83% and 65%, respectively, of the infused volume had moved across the capillary membranes during the 2 hours after volume infusion. Significantly, in all three groups after 120 min, cardiac output (panel B), mean systemic pressure (panel C), and arterial pressure (not shown), all had returned to their initial levels, even though blood volume remained 17% above normal in the whole blood group and 35% above the initial level in the dextran group. These data suggested that, in the two groups with persistent blood volume expansion, the normalization of mean systemic pressure was achieved by an expansion of vascular capacity sufficient to contain the greater volume at the initial level of pressure. The mechanisms responsible for the accommodation

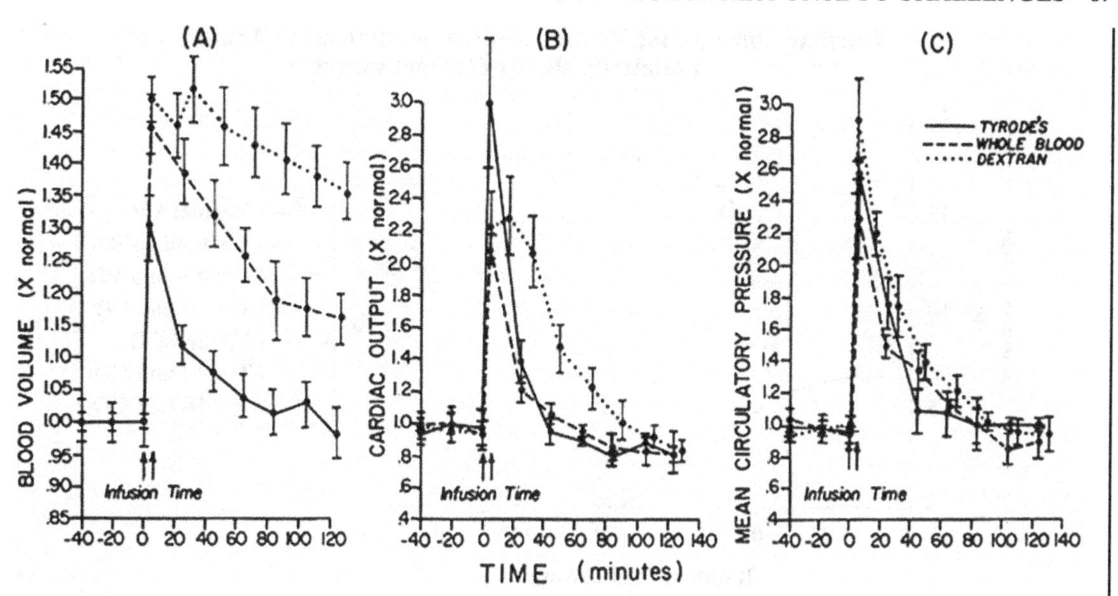

FIGURE 6.5: Data showing response over a 2-hour period following volume expansion with Tyrode's solution (solid lines), whole blood (dashed line), and saline solution (dotted line). (A) Blood volume data, (B) cardiac output, and (C) mean circulatory filling pressure. Reproduced with permission of Elsevier from reference [47], derived from data from reference [46].

may have been stress relaxation of vascular tissue and enlargement of capacities of reservoir structures. However, while reflexively mediated withdrawal of sympathetic stimulation could not have been elicited in response to an increase in arterial pressure, which was normal at the 120-min point, low pressure receptors in the atria may have been activated (atrial pressure or central venous pressure were not reported) and signaled a reduction in sympathetic tone that could have contributed to the return of mean systemic pressure to the normal level.

The sequence of events in the response to blood volume expansion can be illustrated and analyzed graphically, as shown in Figure 6.6. The normal cardiac function and venous return curves are presented as black lines.

With sudden volume expansion of approximately 20%, the immediate response of the vasculature is an increase in mean systemic pressure to 16 mm Hg and a reduction in resistance to venous return due to enlargement of the capillaries and veins, resulting in a rightward rotated and shifted venous return curve, shown as the purple solid line. The theoretical immediate cardiac function curve may not be affected since arterial pressure may not rise at the "immediate" time point. If that were the case, the altered venous return curve would intersect the unchanged cardiac function curve at 17 L/min and right atrial pressure of 4 mm Hg. Within a few seconds, autonomic reflexes would

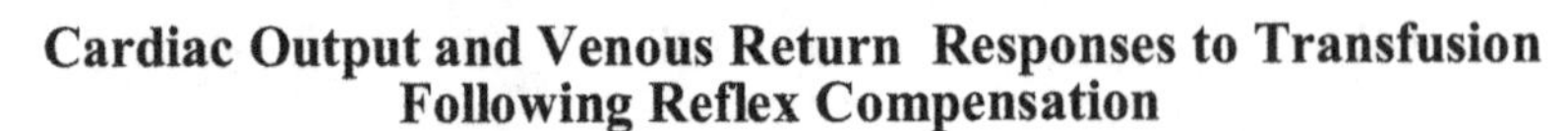

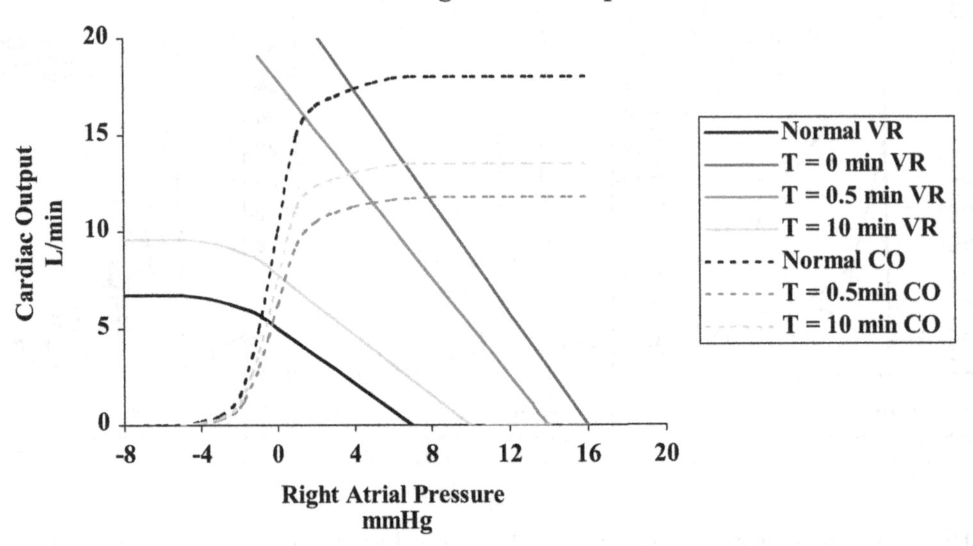

FIGURE 6.6: Expected responses to a rapid transfusion of blood on venous return and cardiac output. The immediate responses and those occurring at 0.5 and 10 min after transfusion are included.

alter the cardiac function curve, reducing its slope and plateau, as illustrated by the blue dashed curve. Sympathetic withdrawal would also affect the venous return curve within seconds (solid blue line), reducing mean systemic pressure to approximately 14 mm Hg. The two function curves describing the cardiovascular system at this time point would intersect at 11 L/min cardiac output and atrial pressure of 5 mm Hg. After approximately 10 min, stress relaxation would continue to affect the venous return curve, reducing mean systemic pressure to 10 mm Hg. With the reduction in pressure, resistance to venous return would increase, yielding the venous return curve shown as the yellow line. Cardiac output and arterial pressure would progressively decrease over 10 min, allowing sympathetic stimulation of heart to increase toward the normal level, shifting the cardiac function curve toward the normal position, shown as the yellow dotted line. The two function curves would intersect at 7 L/min cardiac output and approximately 0 mm Hg right atrial pressure.

6.3 CIRCULATORY SHOCK

Many definitions have been proposed for the term circulatory shock, none of them adequate for all conditions commonly referred to by that label. A broad definition that is generally applicable is that circulatory shock is the condition in which cardiac output is insufficient to provide adequate oxygen and nutrition to the tissues of the body, causing progressive, global cellular damage. The condition can result from several causes, including peripheral factors, such as blood loss, dehydration, and loss of vascular tone, and cardiac factors, such as myocardial infarction, progressive myocardial ischemic disease, valvular failure, cardiac tamponade, and cardiomyopathy.

Peripheral factors leading to circulatory shock cause mean systemic pressure to fall, or rarely, resistance to venous return to increase. Reductions in blood volume due to hemorrhage or dehydration can reduce mean systemic pressure and consequently the pressure gradient for venous return to levels that cannot support adequate cardiac output. If cardiac output remains inadequate for prolonged periods, tissue damage and circulatory shock will develop.

Even with normal blood volume, extreme vasodilation throughout the body such as that associated with anaphylaxis can have the same effect, causing an increase in systemic capacitance and/or unstressed vascular volume to a degree that reduces mean systemic pressure and consequently pressure gradient for venous return to critically low levels. Numerous factors have been identified as being responsible for the widespread vascular effects of anaphylaxis, most being cytokines and oxygen-free radicals including nitric oxide originating from white blood cells and vascular endothelial cells released in response to the substance initiating the crisis.

Elevation of resistance to venous return can also impair return of blood to the heart to a critically low rate. Prolonged positive pressure respiration at levels high enough to significantly elevate pressure in the thorax can compress veins leading to the heart and greatly increase resistance. Surgical procedures in the thorax or abdomen may also inadvertently compress one or more of the great veins. Abdominal masses such as tumors, especially those associated with the liver, can gradually compress the inferior vena cava, elevating resistance to venous return. Accumulation of ascites fluid within the abdominal cavity may compress all structures in the abdomen, increasing venous resistance.

Cardiac pathology or damage may reduce cardiac output to levels that lead to circulatory shock if output is unable to provide sufficient oxygen and nutrients for cellular metabolism. Such conditions are referred to as cardiogenic shock.

In shock of any etiology, if blood pressure falls to sufficiently low levels, the myocardium itself will become underperfused relative to its requirements, with the result being the weakening of the pumping ability of the heart and a further reduction in cardiac output. Therefore, in any severe case of circulatory shock, the weakening heart itself becomes a causative factor contributing to the condition.

Compensatory mechanisms respond rapidly when blood volume falls or resistance to venous return rises to a degree that significantly impairs cardiac output or reduce arterial blood pressure. The same mechanisms discussed earlier in the chapter act in ways that limit the decline in arterial pressure and cardiac output, beginning within seconds or a few heartbeats of a fall in blood pressure. The sympathetic nervous system elicits reflex-mediated contraction of the capacitance vessels and structures, offsetting the effect of blood volume loss on mean systemic pressure. At the same time, increased sympathetic stimulation of the heart strengthens the myocardium. Reverse stress relaxation of the capacitance vessels may also act to attenuate reductions in mean systemic pressure during blood volume deficits. Reduced capillary pressure due to decreased flow alters the Starling forces, creating transudation of fluid into the vascular space, raising blood volume toward the initial level.

Several hormonal systems are also recruited to assist in cardiovascular support during circulatory shock. The renin–angiotensin system is stimulated, first by sympathetic reflexes initiated by reductions in left and right atrial pressure, then by reflexes arising from the arterial baroreceptors, and ultimately by the effects of arterial hypotension directly on the renin release mechanisms in the afferent arterioles of the renal cortex. The result of these actions is an increase in the rate of renin release and a consequent elevation in the concentration of angiotensin II in the blood. Angiotensin II is a powerful vasoconstrictor not only acting prominently on the precapillary resistance vessels but also affecting the capacitance vessels, significantly increasing mean systemic pressure [5]. Together, these actions have a significant effect on arterial blood pressure recovery from severe hemorrhage [47].

Vasopressin release is strongly stimulated by reduction in left atrial pressure and arterial pressure as well. The reflex release mediated in these ways gives rise to much greater rates of release of the hormone than is needed for osmotic regulation, yielding concentrations in the blood that have significant effects on arterial pressure recovery, via actions on both arterial resistance and on systemic capacitance [48].

The arterial baroreceptor initiated sympathetic reflexes, and possibly, those arising from the atrial receptors stimulate adrenal medullary secretion of epinephrine. This catecholamine has significant positive cardiac ionotropic effects and constrictor effects on the capacitance structures and vessels, increasing mean systemic pressure.

In severe circulatory shock, if the arterial blood pressure falls below 50 mm Hg, the central nervous system ischemic reflex will be elicited, driving all reflex-mediated compensatory mechanisms to their maximal responses.

Progressive shock develops when damaging positive feedback systems overwhelm the compensatory negative feedback actions. The differences between recovering circulatory shock and progressive shock can be considered in terms of feedback control. A negative feedback control system is one in which an error or deviation in the value of the regulated variable, for example, arterial blood pressure, in one direction elicits a response of the control system in the opposite direction, partially or completely correcting the error. If arterial pressure falls due to hemorrhage, the arterial baroreceptor reflex will increase total peripheral resistance and cardiac output, raising arterial pressure back toward the initial level. With each cycle through the system's operation, the compensatory actions of the negative feedback system will reduce the error, improving the condition. Positive feedback systems work in the opposite manner; if the value of the controlled variable changes, the positive feedback system will respond in ways that change the variable even more in the same direction as the initial error. If blood pressure decreases as a result of hemorrhage and if positive feedback systems are operating, their effects will be to cause additional reductions in blood pressure. With each iteration of the cycle, the effect of the positive feedback will add to the previous error, increasing the severity of the condition. Much of normal physiology can be understood in terms of negative feedback control, and much of pathology and disease can be viewed as positive feedback mechanisms, also termed vicious cycles. A limited number of positive feedback systems operate within normal physiological systems. However, in most conditions, negative feedback is virtuous, but positive feedback may be considered a sin, and the wages of sin are disease and death.

Within a limited range of reductions in cardiac output and arterial blood pressure, the gain, or effectiveness, of the negative feedback compensatory mechanisms is greater than the gain of the positive feedback effects resulting from deterioration of the heart and other organs due to inadequate blood flow. Within this range of initial decline, the cardiovascular condition will recover, and the condition is termed recovering shock.

However, beyond a certain level of cardiac output or arterial pressure reduction, the intensity of the damage to the cardiovascular system becomes great enough to increase the gain of the deleterious positive feedback cycles to values exceeding the combined negative feedback gains of the compensatory systems. Once that critical point is exceeded, the positive feedback effect to increase the severity of the initial error with each passing minute will lead to a situation of accelerating decline toward death. This condition is referred to as progressive shock.

The exact tipping point of blood pressure or cardiac output is not known, and undoubtedly, the duration of the hypotension or reduced perfusion is also a critical factor in determining the direction of the prognosis; the "golden hour," 60 min immediately after hemorrhage, is believed by trauma physicians to be the critical period during which aggressive measures to restore circulation can provide a positive outcome, even if the initial condition is extremely compromised. However, a study by Crowell and Guyton [49] illustrates the fine line of balance between positive and negative feedback systems in determining the direction of progression in hemorrhagic shock. In a large study involving six groups of anesthetized dogs, rapid hemorrhage was conducted under controlled conditions to predetermined arterial pressure levels: 80, 63, 49, 43, 35, and 17 mm Hg. The animals were then observed for the next 6 hours. All animals bled to pressure levels of 47 mm Hg or higher recovered and survived, while all bled to levels of 43 mm Hg or below died over the course of the next several hours. At least for these experimental conditions involving rapid hemorrhage during anesthesia, the critical arterial pressure when the deleterious effects of hypotension and inadequate perfusion overwhelm the compensatory mechanisms' ability to respond lies between 49 and

43 mm Hg. Below this level of post-hemorrhage hypotension, the positive feedbacks causing progressive damage to the heart and other critical areas of the circulatory system exceed the capabilities of the opposing compensatory negative feedback systems to correct the pathology.

The exact causes of the progressive increases in gain of damaging positive feedback cycles with increased severity of circulatory shock are not completely understood, but several sources that undoubtedly contribute are known. As mentioned previously, the effects of myocardial underperfusion at arterial pressure levels below 50–60 mm Hg will lead to progressive weakening of the strength of the heart. Initially, the effect on cardiac output may not be significant since the reduction in pumping strength may be offset by the resulting increase in atrial pressure. However, in time, the damaging effects of underperfusion will become more and more severe, eventually to the point at which cardiac output will be affected. From that time forward, the progressively failing heart and resulting reductions in cardiac output will add a major contribution to the rise in gain of the positive feedback cycles leading to the decline of the system.

Once arterial pressure falls below the autoregulatory range of the brain, inadequate perfusion will rapidly affect neurological function. Impairment of the vasomotor center may become apparent with prolonged underperfusion. The result will be a generalized loss of vasomotor tone, reduction in vascular resistance, increased transudation of fluid from the blood across the capillary membranes throughput the body, reduced mean systemic pressure, and large negative effects on cardiac output and arterial blood pressure. All theses effects will contribute to the progressively rising positive feedback gains of the damaging vicious cycles of progressive shock. Eventually, the respiratory centers fail as well.

Many other factors contribute to the accelerating decline characterizing progressive shock, most not as powerful as the effects of the failing heart and neurological centers. The vascular endothelium suffers damage from underperfusion either as a direct metabolic consequence or due to reactive oxygen species released from activated white blood cells. The impacted endothelial cells release additional cytokines, many of which have vasodilator effects or stimulate white cell activity. With extreme states of low cardiac output and tissue underperfusion, some capillaries may experience prolonged periods when flow ceases completely. The stasis together with the damaged endothelium leads to platelet aggregation, initially as platelet clumping that can disaggregate if flow resumes. However, with extended periods of stasis, clotting will develop, permanently occluding the vessel. Crowell termed this process, which can develop throughout the body, *diffuse intravascular coagulation* [50, 51]. It may result in lasting deficits in functions of the brain and other organs even if the patient does not succumb.

If the progressive phase of shock continues for an extended period, irreversible shock will develop, a condition that cannot be corrected regardless of therapeutic attempts. The transition point between progressive but still reversible and irreversible shock is correlated with the total oxygen debt (the difference between the basal oxygen consumption rate and that measured during the period following hemorrhage) accumulated by the patient during the period of reduced cardiac output. In experimental animals in which oxygen consumption was measured from the onset of hemorrhage, Crowell and Smith [52] found that, if the total oxygen debt exceeded 150 mL of oxygen per kilogram, all animals went into the irreversible stage, but if the oxygen debt was less than 100 mL/kg, all animals could recover. Jones [54] reported that the LD_{50} value of oxygen debt was 120 mL/kg and that it was independent of the total time required for the oxygen deficit to develop.

Closely correlated with severe oxygen debt is depletion of cellular adenosine and high-energy adenosine compounds such as adenosine triphosphate (ATP). During periods of inadequate oxygen supply to the heart and other tissues, cells do not have adequate ability to generate high-energy phosphate bonds through oxidative metabolism due to low oxygen levels or through glycolysis due to inhibition of glycolitic enzymes by low pH. Consequently, the cells deplete all high-energy compounds available, first ATP, then adenosine diphosphate, finally adenosine monophosphate (AMP). The adenosine formed from the hydrolysis of AMP readily diffuses out of the cells and is converted to uric acid that cannot reenter the cells. Adenosine can be synthesized within cells but only at a rate of about 2%/hour under conditions of normal metabolic activity. If cells of critically important organs, such as the heart, become depleted of adenosine during a period of prolonged underperfusion, they will be unable to restore their adenosine supply for many hours. Cardiac adenosine depletion can be a significant contributor to development of irreversible shock [54].

Even massive transfusions of blood during the irreversible phase cannot salvage the patient, even though it may temporarily improve cardiac output. Mean systemic and atrial pressures may be elevated to very high levels by infusion of large volumes of blood, yet cardiac output may rise for only a short period, suggesting that the heart, not the peripheral vasculature, is the element of the system responsible for the ultimate failure. Crowell and Guyton repeatedly recorded cardiac function curves during development of irreversible shock in anesthetized dogs. Over the course of the 6-hour study, cardiac function declined until the maximum output was reduced to approximately 20% of normal, even at atrial pressure levels greater than 10 mm Hg [55]. As the time during which the heart was underperfused progressed, the myocardium suffered more and more damage, eventually reaching a point at which its condition was irreversible. Therapy directed to improve cardiac function can delay the point at which shock becomes irreversible. But even with all available interventions, there is a degree of cardiac impairment that can accumulate during prolonged circulatory shock that precludes recovery.

Figure 6.7 illustrates the effects of hemorrhage to irreversible shock on the venous return and cardiac function curves. The normal curves drawn in the black lines intersect at a cardiac output/venous return value of 5 L/min and right atrial pressure of 0 mm Hg. A sudden hemorrhage of approximately 12% of the total blood volume reduces mean systemic pressure to 1 mm Hg and

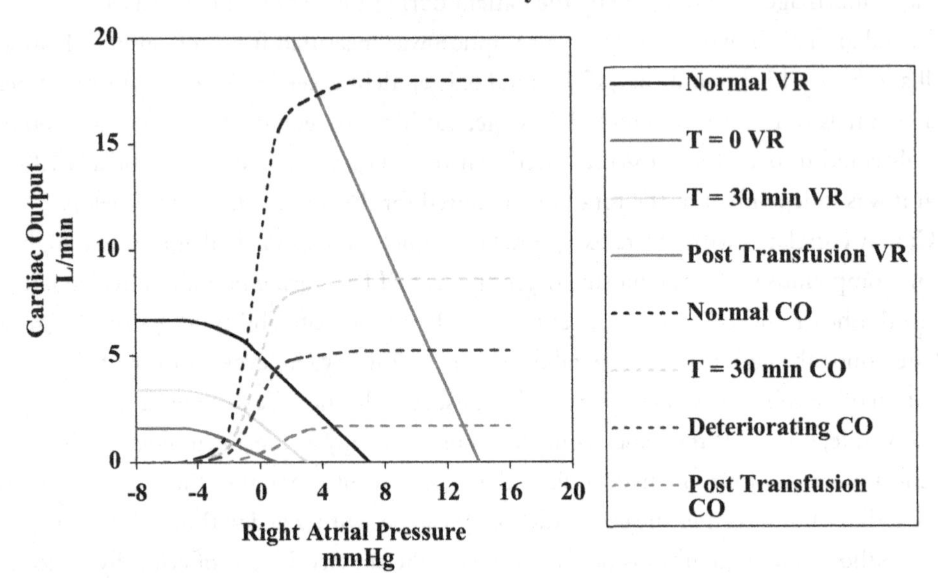

FIGURE 6.7: Expected changes over a period of several hours in venous return and cardiac function after a severe hemorrhage of 12% of the blood volume resulting in progressive and irreversible circulatory shock.

shifts the venous return curve to that illustrated by the purple line, intersecting the momentarily unchanged cardiac function curve at 0.7 L/min and right atrial pressure of −2 mm Hg. After approximately 30 min, compensatory mechanisms respond with reverse stress relaxation, increased sympathetic nervous system reflex stimulation of the heart and vasculature, and fluid movement into the capillaries from the interstium throughout the body. As a result, the venous return curve and cardiac function curves are shifted to those illustrated as the yellow lines. The cardiac function curve has moved to the left, whereas the venous return curve is shifted upward by the increase in mean systemic pressure, resulting in an intersection at a cardiac output value of 2.4 L/min and right atrial pressure of −1.0 mm Hg. The patient might remain in this condition for several hours; however, if cardiac output does remain at this level of <50% of normal, the deleterious effects of underperfusion of the heart and other organs will begin to cause more and more serious damage to the myocardium, weakening the heart and shifting the cardiac function curve to the right and downward, as illustrated by the curve shown as the red line. The equilibrium point now with the unchanged venous return curve shown in yellow is at a reduced value of 1.9 L/min and right atrial pressure of −0.2 mm Hg. From this condition, the system will deteriorate rapidly to the function curve shown as the blue line. If a massive transfusion of blood is given at this point, raising mean systemic pressure to 14 mm Hg, the heart retains insufficient strength to significantly increase output, even at extremely high atrial pressure. The new equilibrium point is at 1.7 L/min and right atrial pressure of 13 mm Hg. The condition of the cardiovascular system will continue to deteriorate at an accelerating rate from this point until death.

However, if the transfusion of blood had been given earlier, 30 min after hemorrhage, when the patient was in the progressive, but not yet irreversible, phase and the heart's condition was reflected in the function curve drawn as the yellow line, the venous return curve shifted by the transfusion would have intersected the cardiac function curve at 8.6 L/min, providing sufficient perfusion to the myocardium and the rest of the body to permit recovery.

6.4 HEART FAILURE

If the heart is unable to pump enough blood to meet the demands of the tissues of the body, the condition is referred to as heart failure. The causes can be related to coronary artery occlusion with infarction of portions of the myocardium, myocardial ischemia associated with diffuse small artery disease throughout the myocardium, cardiomyopathy, valvular insufficiency or restriction, or cardiac tamponade due to fluid accumulation in the pericardium. Failure to pump a sufficient rate of blood flow may be associated with inadequate contractile force or with inability of the myocardium to relax adequately during diastole. The condition can develop suddenly after coronary artery thrombosis or over a period of months or years due to progressive myocardial ischemia or cardiomyopathy. The causes of the condition cannot be considered here, but the responses of the cardiovascular system to heart failure will be analyzed in detail.

In response to an acute reduction in heart strength after an acute coronary artery occlusion, sympathetic nervous system reflexes act within seconds to increase strength of contraction and heart rate and increase mean systemic pressure. The reflexes are initiated initially by reductions in arterial pressure sensed by the arterial baroreceptors, and if blood pressure falls below 50–60 mm Hg, the central nervous system ischemic reflex is activated as well. The reflexes shift the cardiac function curve to the left and upward, whereas the increase in mean systemic pressure shifts the venous return curve to the right. Figure 6.8 depicts the normal function curves, drawn in black, the cardiac function curve at the instant of the event, causing moderate bilateral impairment of heart function (green) such as a coronary artery occlusion, the function curves after 30 s when the rapidly acting reflex compensatory mechanisms have acted (pink), and cardiac and venous return functions after the effects of long-term compensatory functions have responded (yellow), which will be considered in the next paragraph.

Long-term compensation for heart failure includes renal sodium and water retention, which increases extracellular fluid and blood volume. The sympathetic nervous system reflexes act in the kidneys to reduce renal blood flow and glomerular filtration rate, reducing sodium and water excretion as well. In addition, the sympathetic innervation of the afferent arterioles and juxtaglomerular

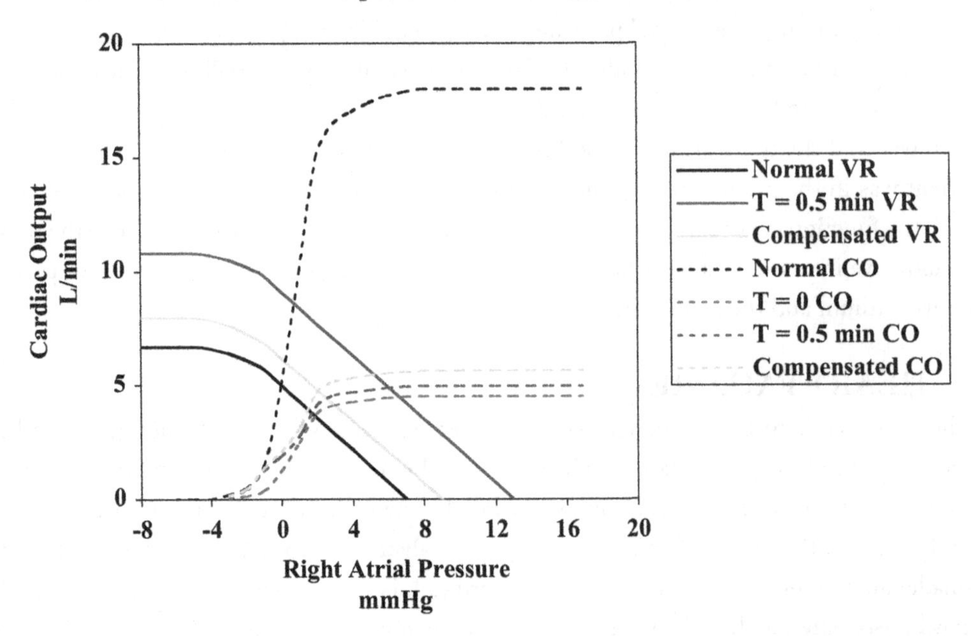

FIGURE 6.8: Expected responses of venous return and cardiac function to an acute coronary artery occlusion resulting in compensated heart failure. Shown are responses at the moment of the event, at 0.5 min, and after compensation.

apparatus of the renal cortex stimulate renin release. The elevated renin levels generate higher concentrations of angiotensin I and II. Angiotensin II, in addition to being a powerful vasoconstrictor that significantly contributes to elevation of blood pressure and mean systemic pressure in the minutes following an acute cardiac event, also has a strong antinatruetic effect, increasing tubular reabsorption of sodium and water. Furthermore, angiotensin II stimulates aldosterone secretion from the adrenal cortex, which also has a significant sodium-retaining effect on the renal tubules.

While the sympathetic reflexes, angiotensin II, and aldosterone all are significant sodium- and water-retaining factors that act on the kidneys after acute cardiac impairment, reduction in arterial pressure associated with reduced cardiac output is the most powerful antinatriuretic influence. Even reductions of a few mm Hg in renal perfusion pressure reduce sodium and water excretion significantly [35, 56–58]. The combination of all sodium- and water-retaining mechanisms acting simultaneously can strongly reduce excretion even down to levels that approach complete retention in severe heart failure.

The yellow curves in Figure 6.8 illustrate the function curves after several days of compensation, when fluid retention has increased blood volume and mean systemic pressure, shifting the venous return curve to the right. In addition, during the days and months following a moderate myocardial infarction, the myocardium can repair and remodel itself in ways that increase the heart's effectiveness as a pump. The beginnings of this process are reflected in the upward shift of the cardiac function curve drawn as the yellow line.

If the responses to the initial impairment of cardiac pumping ability return cardiac output to near the normal resting level, the condition is referred to as compensated heart failure. This is the situation depicted by the yellow function curves intersecting at a cardiac output value of 4.8 L/min, only slightly less than the normal level of cardiac output, although right atrial pressure is significantly greater than normal (2.4 mm Hg). In this condition, the patient has a limited ability to increase his metabolic rate and demand for cardiac output since the plateau of his function curve is only slightly greater than his resting cardiac output. However, since he is able to maintain enough cardiac output to provide for resting conditions and limited exertion, as long as he does not place additional demands on his heart, he will be comfortable. The condition is characterized by stable fluid and electrolyte balance with little or no edema, near-normal blood pressure and blood gases under sedentary, resting conditions, but little cardiac reserve. In addition, his right atrial pressure, mean systemic pressure, blood volume, and extracellular fluid volume will remain elevated, but stable.

Decompensated heart failure is the condition that results when cardiac output remains below the level required to maintain adequate blood flow to the body at rest. Heart failure may progress gradually to the decompensated condition in many chronic cardiac conditions, or it may develop suddenly as a consequence of a severe myocardial infarction associated with a major coronary artery occlusion. Regardless of the cause, if cardiac output is significantly below normal for an extended period after the sympathetic reflexes and hormonal systems have achieved their maximum compensatory effects, fluid and electrolyte excretion will be severely restricted resulting in cumulative positive fluid and electrolyte balances. The sodium- and water-retaining mechanisms described above continue to act on the kidneys as long as cardiac output is below the minimum level required for normal tissue metabolism. Consequently, persistence of the condition for several days or more leads to a large positive fluid balance that may give rise to peripheral edema, which is free fluid accumulation in the tissues. If right atrial pressure rises excessively, free fluid may also collect in the abdominal cavity as ascites. Predominant left-sided heart failure may lead to left atrial pressure levels high enough to produce pulmonary edema before peripheral edema forms, although pulmonary edema may occur in severe bilateral failure as well. Decompensated failure is an unstable condition that may deteriorate progressively toward death unless intervention can salvage the pumping ability of the heart.

Sustained positive fluid and electrolyte balance in heart failure can expand blood volume to levels that result in very high right atrial pressure, approaching even 15 mm Hg. Associated with these very high levels of cardiac preload is a progressive deterioration of cardiac function. As the atrial pressure increases, ventricular end diastolic radius increases, which, according to the law of Laplace, increases ventricular wall stress in proportion to the increase in radius. To develop pressure to eject blood from the ventricle, the myocardium must perform work to overcome wall stress. In severe heart failure, the heart has insufficient metabolic energy to perform the workload required of it and suffers progressive damage as a result. When even additional demands associated with severely elevated atrial pressure and consequent ventricular dilation are imposed, further metabolic damage may result, thus worsening the heart's pumping ability. The critical situation associated with high atrial pressures is undoubtedly exacerbated by the concomitant occurrence of inadequate coronary blood flow and arterial hypoxia that may develop from pulmonary edema. Clearly, such a condition has the features of a positive feedback cycle that can accelerate into a rapidly fatal vicious cycle.

The progression of decompensated failure can be analyzed graphically, as presented in Figure 6.9. The normal function curves are drawn in black lines. In this example, the patient suffers a serious coronary artery occlusion that suddenly shifts his cardiac function curve downward and to the right (drawn in red), reaching an intersection with the momentarily unchanged venous return curve at 1.0 L/min cardiac output and right atrial pressure of 6 mm Hg. With compensation from the reflex mechanisms, after approximately 10 min, the cardiac function curve may be improved to that illustrated in purple that reaches an equilibrium with the compensated venous return curve at 2.0 L/min and right atrial pressure of 7 mm Hg. Over the ensuing hours, cardiac function remains stable but cannot improve significantly, although fluid retention begins to increase blood volume and mean systemic pressure, shifting the venous return curve to the right (shown in yellow). Although the new equilibrium point is on the same cardiac function curve, cardiac output is now 2.5 L/min due to the increase in right atrial pressure (9 mm Hg). Several more days of positive fluid balance continues to shift the venous return curve to the right, although the cardiac function curve remains unchanged. Once the venous return curve has shifted rightward to a degree that the intersection with the cardiac function curve is on the plateau (blue venous return curve) at a cardiac output 3.0 L/min and right atrial pressure of 15 mm Hg, additional fluid retention, blood volume increases, and right atrial pressure elevations cannot elicit additional increases in cardiac output. However, fluid retention continues, shifting the venous return curve and raising right atrial pressure still higher (green venous return curve); as a result, the heart weakens, shifting the cardiac function curve downward and to the right (green cardiac function curve) so that the new equilibrium value is at cardiac output of 2.6 L/min and right atrial pressure of 18 mm Hg. Fluid retention continues unabated, even more avidly as arterial pressure falls, shifting the venous return curve farther to the right, raising atrial pressure, further weakening the heart so that the subsequent equilibrium points are at progressively lower values of cardiac output and higher right atrial pressures. The cycle of deterioration usually progresses rapidly to death.

Treatment of heart failure focuses on shifting the cardiac function curve upward and to the left and shifting the venous return curve to the left to reduce atrial pressure. Cardiac function can be strengthened by a variety of pharmacological means that have specific positive ionotropic effects on

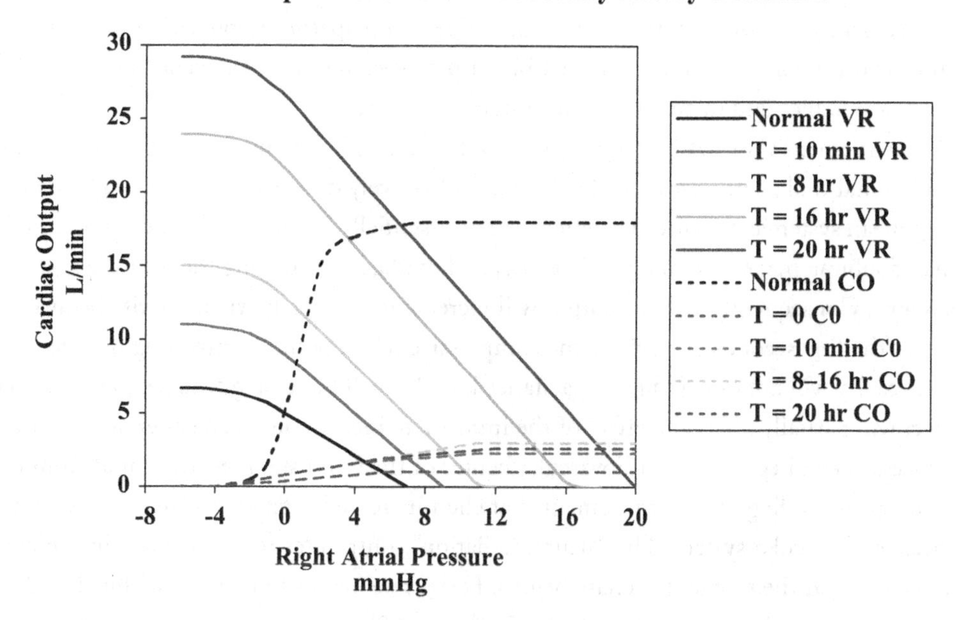

FIGURE 6.9: Responses expect in venous return and cardiac function following a severe coronary artery occlusion leading to decompansated heart failure are illustrated in the figure. The immediate change and subsequent responses over the next 20 hours are presented.

the myocardium. Antihypertensive therapy is also useful in less severe conditions of failure because the reduction in blood pressure can significantly improve performance of a weakened heart. Reducing the work required of the heart by restricting activity of the patient is beneficial to patients who have the potential to improve cardiac function if they have an opportunity to repair or remodel the myocardium following damage. Diuretics provide benefits to patients with elevated atrial pressure by overcoming the sodium- and water-retaining mechanisms elicited in response to reduced cardiac output.

6.5 EXERCISE

Nearly all aspects of cardiovascular function engage in the responses to vigorous physical exercise. Beginning with the first heartbeat after the start of strenuous exertion, the factors that regulate venous return and the strength of the heart all contribute to the cardiovascular acceleration that can multiply the rate of flow through the system by a factor of 3 or 4 (much higher in trained athletes) within a minute or two.

The autonomic effects on the cardiovascular system often begin before the start of exercise. If the subject is aware of the approach of an event or condition demanding physical exertion, her autonomic nervous system will begin to increase sympathetic drive to the heart and vascular system and decrease parasympathetic traffic to the heart. These anticipatory responses increase heart rate, strength of cardiac contraction, and arterial blood pressure, and alter the venous return curve by decreasing capacitance and increasing mean systemic pressure.

At the moment vigorous exercise begins, contraction of large muscle groups and the muscles of the abdominal wall compresses the blood vessels, effectively decreasing systemic capacitance and increasing mean systemic pressure. The effect can be essentially instantaneous and quite significant, more than doubling mean systemic pressure [12] and markedly increasing the pressure gradient for venous return. Consequently, cardiac output will increase immediately when activity begins.

After the first seconds of exercise, the compressive forces within contracting muscles and the abdominal cavity will not only reduce capacitance but also will increase resistance to venous return. This effect will partially offset the effect of the immediate increase in mean systemic pressure.

Once exercise begins, the cardiovascular center of the brain stem receives input from central motor centers controlling muscle movement that elicit increased sympathetic nervous system drive to the heart and vascular system. This "neural radiation" contributes to the increase in sympathetic activity occurring at the moment exercise begins. Heart rate begins to increase within 1 to 2 s after the start of exercise and can more than double in the next 20 s.

As exercise progresses, the sympathetic system response continues at high levels and includes epinephrine release from the adrenal medulla. The exact nature of the inputs that stimulate the continuing sympathetic excitation are not well understood, since the usual suspects responsible for sympathetic responses are at normal levels: arterial blood pressure is usually at normal or elevated levels during exercise, arterial oxygen levels are not depressed, carbon dioxide concentration may be less than normal, and, at least during moderate exercise, lactic acid concentration is not elevated. The sustained response may be due to the neural radiation effect or to other unidentified factors.

Although the sympathetic nervous system responses significantly shift the cardiac function curve during exercise upward and to the right, denervation of the hearts of experimental animals only slightly reduces the maximum cardiac output responses during exercise, although the responses are somewhat delayed. Apparently, the Starling effect of increased atrial pressure and ventricular end diastolic volume resulting from shifts in the venous return curve is capable of adequately increasing cardiac output to meet the demands of exercise without input from the autonomic nervous system. Certainly, without sympathetic stimulation, heart rate responses will be severely attenuated and stroke volume will be much greater than normal during strenuous workouts.

Increased metabolic activity of working muscles results in greatly enhanced muscle blood flow during exercise. The metabolic effect is evident within <10 s of the initiation of work and can reach a maximum within several minutes. The effect can be as great as a 15-fold increase in muscle blood flow compared to resting levels. The decrease in vascular resistance of the working muscles more than offsets the increase in resistance to venous return caused by contraction of muscles in the abdominal wall and around the veins within the muscles.

The primary metabolism-related variable responsible for the increase in flow is probably reduced oxygen content of the muscle tissue or some variables closely related to oxygen levels. Other factors that may contribute to the local vasodilation include low pH or elevated concentrations of lactic acid, potassium, adenosine, or other metabolites. Changes in metabolism-related variables may act through release of nitric oxide, prostacyclin, or other vasodilatory locally acting mediators from vascular endothelial cells. In addition, there is evidence that beta adrenergically mediated vasodilator mechanisms contribute to the initial muscle vasodilation occurring at the start of exercise.

Cardiac output and maximal oxygen consumption by the body ($\dot{V}O_{2max}$) are closely correlated. Figure 1.1 illustrates data from studies in which maximal cardiac output and $\dot{V}O_{2max}$ were recorded in a group of normal subjects as well as in trained athletes. The correlation was very close among all individuals, suggesting that metabolic demand for oxygen by the entire body affects resistance to venous return and, by doing so, is a primary determinant of cardiac output. It also suggests that the maximal rate of $\dot{V}O_{2max}$ is determined by the maximum ability of the cardiovascular system to supply the metabolic demands of the tissues. Other evidence that maximal cardiac output determines $\dot{V}O_{2max}$ will be presented in later paragraphs of this section.

In maximally strenuous aerobic exercise, the cardiovascular system responds to produce the highest possible rate of cardiac output. The responses can be analyzed graphically, as in Figure 6.10. The normal function curves are presented in black. In maximal exercise, the cardiac function curve is affected by maximal sympathetic nervous system stimulation as well as by high levels of epinephrine released from the adrenal medulla, shifting the curve (shown in red) to the left of the normal curve. The new curve not only has a greater slope but also has a much higher plateau. The new venous return curve shown in red is shifted to the right and rotated upward by the increase in mean systemic pressure and the reduction in resistance to venous return. The combined effects of abdominal compression and sympathetic stimulation of the capacitance vessels and liver and spleen raises mean systemic pressure to the highest level possible under physiological condition, more than 20 mm Hg. The extreme vasodilation in large working muscle groups contributes to a reduction in resistance to venous return, increasing the slope of the function curve. As a result of these powerful effects on both the cardiac function and venous return curves, the equilibrium point for the system is at 20 L/min and 2 mm Hg, the highest physiological level of cardiac output in this normal untrained subject.

The illustration indicates that the equilibrium point is not at the highest point of the cardiac function curve, not on the plateau portion. Whether or not the maximum output in exercise is on the plateau of the function curve will vary among different subjects and within subjects depending

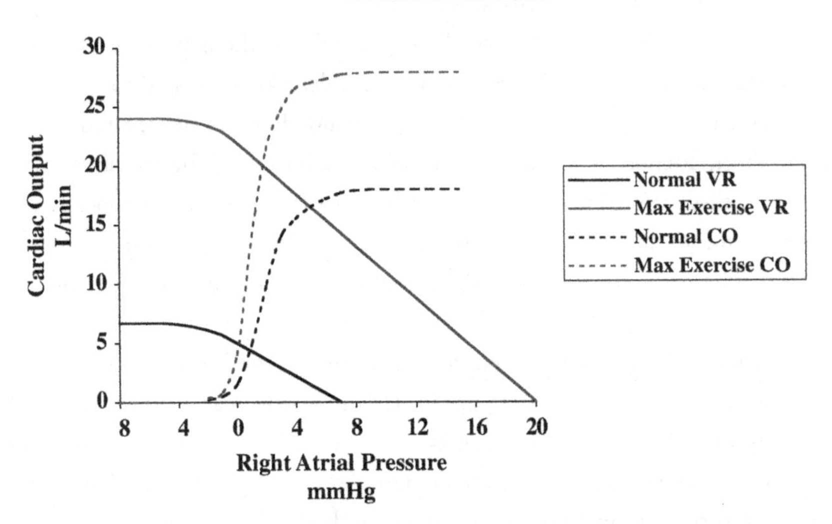

FIGURE 6.10: Venous return and cardiac function responses to maximal exercise.

on several factors. One of the most important considerations is the blood volume of the individual at the time of peak exertion. Especially during condition when the systemic capacitance vessels are maximally stimulated by sympathetic stimulation, mean systemic pressure will be very sensitive to changes in blood volume, and consequently, the position of the venous return curve may be significantly shifted by relatively small changes in blood volume. If the blood volume is increased by transfusion just before the period of exertion (e.g., "blood doping" before the event), the venous return curve may be shifted to the right due to an increase in mean systemic pressure. As a result, the equilibrium point will be shifted to a higher level on the cardiac output function curve. The increase in output enables the individual to sustain a higher level of $\dot{V}O_{2max}$, which is a critical factor in determining performance in aerobic sports. The competitive advantage afforded by blood doping supports the concept that $\dot{V}O_{2max}$ is a function of maximum cardiac output.

Frequently, during long periods of strenuous work or exercise, dehydration due to sweating reduces blood volume and has the opposite effect on the cardiovascular system. Rates of sweating can approach 5 L/h in persons acclimatized to work in hot environments, leading to significant extracellular fluid and blood volume contraction within the time span of a soccer match, football game, or military combat engagement. Mean systemic pressure will be shifted to a lower level, moving the venous return curve to the left so that the equilibrium point will be at a lower point on the cardiac function curve. This shift and consequent reduction in cardiac output and $\dot{V}O_{2max}$ are some of the causes of reduced maximal performance associated with dehydration.

Also negatively impacting maximal performance in warm temperature environments is the demand for blood flow to the skin for thermoregulation. With elevations of core body temperature of several degrees Celsius, not uncommon in extend periods of maximal exertion in warm surrounding, body surface blood flow may increase to as much as 5 L/min, which represents approximately one-third of the maximum increase in cardiac output available in a normal adult, leaving one-third less to support increased metabolic demand.

The cardiovascular system undergoes adaptation in response to aerobic physical training. One of the first effects of aerobic training is an increase in plasma volume. Even during and after the first training session, cortisol secretion from the adrenal cortex is elevated, which stimulates hepatic production of albumin, the principal protein of the plasma. Consequently, a measurable increase in plasma volume in the absence of an increase in total extracelllular fluid volume can be observed within 24 hours of the beginning of training [59]. After three to six training sessions, plasma volume increases to between 12 and 20%, the increase being in proportion to the increase in total plasma protein [60, 61]. Plasma volume returns to normal within 1 week of cessation of training [62].

With longer periods of training, changes take place in the vascularity of muscles involved in the specific activities of the training program. Arteries supplying the affected muscles enlarge in radius, and capillaries increase in density. Over periods of weeks, these adaptations become large enough to have significant effects on distribution of blood flow during exercise, with a greater portion of total cardiac output going to the working muscles due to their reduced vascular resistance. This accounts at least in part for the observation that, after a period of training, subjects can perform work at a given rate of oxygen consumption with a lower cardiac output than before training [63]. In one study, work at specified levels of oxygen consumption required 1.1–1.5 L/min less total cardiac output after 16 weeks of training, even though maximal cardiac output was capable of increasing from 22.4 to 24.2 L/min [64].

The pumping ability of the heart improves with training due to increases in mass of the ventricular walls and in left ventricular cavity volume [65]. Modest cardiac hypertrophy occurs with long-term training, but heart mass returns to normal after training is discontinued. Left ventricular end diastolic volume increases both at rest and during exertion. As a result, resting and maximal heart rates are lower after a period of aerobic training. In highly trained competitive athletes, end diastolic volume is as much as 25% greater than in untrained subjects; however, the difference is probably more related to genetically determined differences in cardiac dimensions than to training effects.

These and other adaptations contribute to the increased cardiovascular performance that is associated with aerobic training. Maximum cardiac output improves with periods of training to degrees that vary among studies depending on the condition of the subjects before training and the intensity and duration of training; the range of improvement observed in numerous studies may

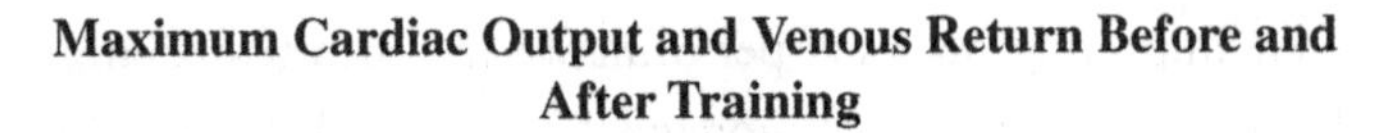

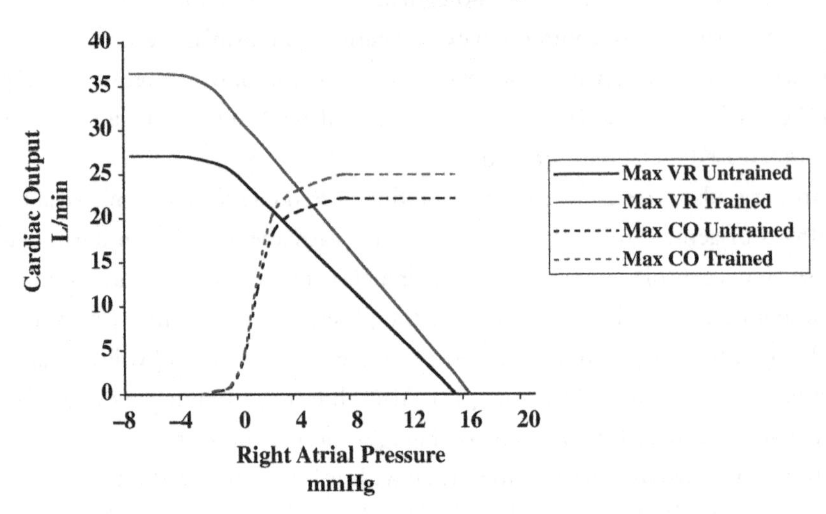

FIGURE 6.11: Expected effects in a normal subject of several months of aerobic training on cardiac function and venous return during maximal exercise.

extend from 8 to 12%. Enhancement of stroke volume accounts for the additional cardiac output. Improvement of blood flow distribution during exercise together with the improvement in maximal cardiac output combine to provide for increases in $\dot{V}O_{2max}$ of as much as 30% in normal subjects in response to weeks of training.

The effects of training on the venous return and cardiac function curves during maximal exertion are illustrated in Figure 6.11, with the normal, maximal pretrained functions drawn in black lines, with the equilibrium value during maximal exercise of 19.0 L/min and right atrial pressure of 2.5 mm Hg. With training, maximal mean systemic pressure during exercise is higher (17 vs 15 mm Hg) due to increased blood volume, and the resistance to venous return is reduced due to increased vascularity in the large muscle groups involved in the exercise. The cardiac function curve is shifted to the left and the plateau is raised as a result of the enlargement and strengthening of the heart. The equilibrium point of the modified, "trained" function curve is at a higher cardiac output/venous return (23.2 L/min) and right atrial pressure at 4 mm Hg.

World class aerobic athletes can achieve much higher rates of cardiac output and $\dot{V}O_{2max}$ than the normal subjects studied in the previously discussed studies. Maximal cardiac outputs approaching 40 L/min are common in the world's best distance runners, swimmers and cyclists. The differences between the cardiovascular systems of such individuals and those of normal subjects are probably related to genetic determination of heart size as well as the extensive training champion athletes undertake.

6.6 SUMMARY

In this chapter, a diversity of unrelated cardiovascular challenges was analyzed. All the conditions were complicated and have been difficult to understand for clinicians and scientists. However, by considering the challenges in terms of effects on the venous return and cardiac function curves, comprehensive analyses of the situations are relatively straightforward and lead to logical understanding of the integrated responses of the cardiovascular system.

• • • •

CHAPTER 7
Conclusion

Guyton's contributions were of great importance, although the works of his associates and others in the field surely were also significant. Several key advances of Guyton and those with whom he worked led the way forward. One of their first critical developments was a technique using extracorporal shunts and pumps to repeatedly measure mean systemic pressure in animals. He combined this with the method he developed for continuous measurement of cardiac output using the Fick principle with oxygen saturation in the arterial and venous blood as the indicator. He developed the equipment for cardiac output measurement more than 50 years ago, long before electromagnetic and Doppler flow meters that have been used routinely in research for the last several decades. His methods made possible determination of complete venous return curves during manipulation of the variables that affect the system. He was able to analyze exactly how changes in blood volume affect the system, the significance of resistance to venous return, the importance of the differences in capacitance in the arterial and venous beds, and the pivotal role of the pressure gradient for venous return in determining the rate of flow to the heart. The knowledge he gained from analyzing venous return led to understanding of other significant cardiovascular concepts, possibly the most far reaching being the dominant long-term role of whole-body tissue oxygen demand in determining cardiac output, and the mechanisms by which tissue regulation of blood flow to meet local metabolic demand for oxygen could affect systemic resistance to venous return for the entire body.

During the same period in the 1950s and 1960s, he perfected preparations that enabled control of atrial pressure over a wide range while cardiac output was measured continuously, yielding complete cardiac output function curves obtained over a period of a few minutes, before excessive demand on the myocardium damaged and degraded function. His subsequent realization that cardiac function and venous return curves intersect at only one point at any given moment, at the equilibrium point, led to decades of fruitful analyses of the operation of the integrated cardiovascular system that remains the basis for understanding the regulation of cardiac output.

In the latter half of the 1960s, Guyton and Coleman began developing mathematical models of the cardiovascular system. Initially, the models were small and limited in scope, but even the first efforts produced new insights into the control of the system. As they expanded the limits of the models, they were able to begin considering the dynamics of the system, moving beyond the single point in time analyses considered using graphical techniques. Once time was included as a variable in the dynamic models, a new level of complexity was introduced into the understanding of cardiovascular control.

The importance of inclusion of time as a variable is intuitively obvious; however, many of the implications were not apparent until the models were developed. In the mid-1960s, Coleman and Guyton were working together with an early version of the digital cardiovascular model running simulations, and they observed a result, a surprising one at the time: regardless of how the cardiovascular variables were manipulated, steady-state arterial blood pressure could not be changed from its initial value. Changes in strength of cardiac contraction (within the range that did not cause heart failure), resistance to venous return, arterial resistance, capacitance of various segments, blood volume, and mean systemic pressure could not affect the level of arterial pressure for longer than a few days.

The explanation they soon discovered was the function relating the value of arterial pressure perfusing the kidneys to the rate of renal excretion of sodium and water. As long as this function is unchanged and intake of sodium and water are constant, changes in other components of the cardiovascular system could not alter the steady-state value of arterial pressure. If an initial change is introduced in a function that increased arterial pressure, for example, a decrease in vascular compliance resulting in an elevation of mean systemic pressure, pressure gradient for venous return, right atrial pressure, and cardiac output, the resulting elevation of renal perfusion pressure would raise the rate of renal excretion of extracellular fluid to a value exceeding the rate of intake. Consequently, extracellular fluid would decrease by a few milliliters with each iteration of the model. With the simulated passage of time, extracellular fluid volume would decrease progressively, reducing blood volume, mean systemic pressure, and the subsequent dependent variables, including arterial blood pressure. Extracellular fluid balance would continue to be negative as long as renal perfusion pressure was above the initial control level. Even if the perfusion level were only slightly greater than normal, the model simulations indicated that the renal excretory–body fluid volume–arterial pressure negative feedback system had the capacity to correct completely an initial error in arterial blood pressure leaving no residual error. This was the case for correcting errors that caused either initial positive or negative arterial blood pressure errors. However, in many situations, the model predicted that several days would be required for the negative feedback system to return the level of arterial pressure to the initial value. The potency of the renal excretory rate–body fluid volume–arterial pressure negative feedback system in blood pressure regulation predicted by the cardiovascular model led to decades of experiments concerning hypertension. Ultimately, a hypothesis concerning the causes and treatment of the disease was developed that was based on changes in the relationship between renal perfusion pressure and excretion of sodium being the cause of all forms of hypertension.

Long-term blood pressure regulation and hypertension are the subject of a forthcoming volume in this series by J. P. Granger.

In the four or five decades since they completed their first series of cardiovascular analyses, Guyton and colleagues and many others have moved forward from the basis they established. However, none of the thousands of later studies disproved or even raised doubts about the validity of the initial work; rather, subsequent investigations have reaffirmed the importance of the logic derived from those studies carried out in Jackson beginning in the 1950s and 1960s. Its tenets can be summarized briefly and generally as follows:

- Cardiac output is always equal to venous return to the heart, except for brief, transient periods.
- The heart's output is determined by the rate of venous return from the peripheral tissues. Within physiological limits, the heart will increase output over a wide range in response to small elevations in the right atrial pressure resulting from increases in venous return. For a normal heart operating in the physiological range, increases in strength of contraction alone will not result in significant, sustained increases in output, unless venous return also increases. In addition, reduction in strength of the heart by disease may not affect the normal resting level of cardiac output if the plateau value of the weakened heart's function curve is greater than the body's demand for flow under normal conditions.
- Venous return is a function of resistance to venous return and the pressure gradient for venous return. Resistance to venous return is the total resistance from the root of the aorta to the right atrium. The pressure gradient for venous return is the difference between mean systemic pressure and right atrial pressure.
- Resistance to venous return is affected by factors that cause changes in smooth muscle tone of resistance vessels or changes in pressure in the tissue surrounding thin-walled venous structures. One of the most powerful factors affecting resistance to venous resistance is autoregulation of blood flow to meet local tissue demand throughout the body for oxygen and other metabolic requirements. Consequently, whole-body oxygen demand and metabolic rate are primary long-term determinants of cardiac output.
- The pressure gradient for venous return is the difference between mean systemic pressure and right atrial pressure. Mean systemic pressure is affected by the capacitance of the circulatory system, its unstressed volume, and the volume of blood circulating within it. Right atrial pressure is a function of the rate of flow into it from the veins and the pumping ability of the ventricles. In conditions of health, the normal heart can increase or decrease output over the range normally required with only a few millimeters of mercury variation in atrial

pressure. Conversely, mean systemic pressure commonly may increase as much as 100% from the resting level in response to physiological demands.

- Cardiac and circulatory system functions both respond to cardiovascular challenges. The sympathetic and parasympathetic nervous systems alter cardiac function by affecting heart rate and strength of contraction, shifting the cardiac function curve either upward and to the left (sympathetic) or downward and to right (parasympathetic). The sympathetic nervous system affects many of the determinants of venous return, including mean systemic pressure and resistance to venous return. Circulation hormones, including angiotensin II, vasopressin, epinephrine, and locally acting hormones, including prostaglandins, endothelins, and numerous cytokines, affect vascular resistance and hence resistance to venous return. Long-term effects of the sympathetic nervous system and the endocrine system can lead to decreased rates of renal sodium excretion, increasing blood volume, mean systemic pressure, and the pressure gradient for venous return.
- The renal sodium excretion–body fluid volume–arterial pressure negative feedback system regulates steady-state arterial blood pressure, although several days may be required for arterial pressure to return to the initial level following a perturbation. Although alterations in many cardiovascular functions can initially change cardiac output and arterial pressure, only those that affect the basic renal sodium excretion–body fluid volume–arterial pressure negative feedback system can produce sustained changes in arterial blood pressure.

That logic remains the fundamental basis of understanding of cardiac output control.

• • • •

References

[1] Guyton AC, Jones CE, Coleman TG. Normal cardiac output and its variations. Chap 1. In *Circulatory Physiology: Cardiac Output and its Regulation*. pp. 3–20. WB Saunders, London. 1973.

[2] Saltin B. Physiological effects of physical conditioning. *Med Sci Sports*. 1968; 1: pp. 50–57.

[3] Guyton AC, Armstrong GG, Chipley PL. Pressure–volume curves of the arterial and venous systems in live dogs. *Am J Physiol*. 1956; 184(2): pp. 253–58.

[4] Guyton AC, Polizo D, Armstrong GG. Mean circulatory filling pressure measured immediately after cessation of heart pumping. *Am J Physiol*. 1954; 179(2): pp. 261–67.

[5] Young DB, Murray RH, Bengis RG, Markov AK. Experimental angiotensin II hypertension. *Am J Physiol*. 1980; 239(3): pp. H391–98.

[6] Guyton AC, Lindsey AW, Abernathy B, Richardson T. Venous return at various right atrial pressures and the normal venous return curve. *Am J Physiol*. 1957; 189(3): pp. 609–15.

[7] Guyton AC, Lindsey AW, Kaufmann BN, Abernathy JB. Effect of blood transfusion and hemorrhage on cardiac output and on the venous return curve. *Am J Physiol*. 1958;194(2): pp. 263–67.

[8] Richardson TQ, Stallings JO, Guyton AC. Pressure–volume curves in live, intact dogs. *Am J Physiol*. 1961; 201: pp. 471–74.

[9] Prather JW, Taylor AE, Guyton AC. Effect of blood volume, mean circulatory pressure, and stress relaxation on cardiac output. *Am J Physiol*. 1969; 216(3): pp. 467–72.

[10] Guyton AC, Satterfield JH, Harris JW. Dynamics of central venous resistance with observations on static blood pressure. *Am J Physiol*. 1952; 169(3): pp. 691–99.

[11] Richardson TQ, Fermoso JD. Elevation of mean circulatory filling pressure in dogs with cerebral-ischemia induced hypertension. *J Appl Physiol*. 1964;19: pp. 1133–34.

[12] Guyton AC, Douglas BH, Langston JB, Richardson TQ. Instantaneous increase in mean circulatory pressure and cardiac output at onset of muscular activity. *Circ Res*. 1962; 11: pp. 431–41.

[13] Guyton AC, Sagawa K. Compensations of cardiac output and other circulatory functions in areflex dogs with large A–V fistulas. *Am J Physiol*. 1961; 200: pp. 1157–63.

[14] Patterson SW, Starling EH. On the mechanical factors which determine the output of the ventricles. *J Physiol.* 1914; 48(5): pp. 357–79.

[15] Wiggers CJ, Katz LN. The contour of the ventricular volume curves under different conditions. *Am J Physiol.* 1922; 58: pp. 439–75.

[16] Starling EH. *The Lincare Lecture on the Law of the Heart.* Longman, Green and Company, London. 1918.

[17] Sarnoff SJ, Berglund E. Ventricular function. I. Starling's law of the heart studied by means of simultaneous right and left ventricular function curves in the dog. *Circulation.* 1954; 10(1): pp. 84–93.

[18] Sarnoff SJ. Myocardial contractility as described by ventricular function curves; observations on Starling's law of the heart. *Physiol Rev.* 1955; 35(1): pp. 107–22.

[19] Isaacs JP, Berglund E, Sarnoff SJ. Ventricular function. III. The pathologic physiology of acute cardiac tamponade studied by means of ventricular function curves. *Am Heart J.* 1954; 48(1): pp. 66–76.

[20] Guyton AC, Jones CE, Coleman TG. *Circulatory Physiology: Cardiac Output and its Regulation.* Figure 9–3, p. 161. WB Saunders, London. 1973.

[21] Sagawa K, Ross JM, Guyton AC. Quantitation of cerebral ischemic pressor response in dogs. *Am J Physiol.* 1961; 200: pp. 1164–68.

[22] Herndon CW, Sagawa K. Combined effects of aortic and right atrial pressures on aortic flow. *Am J Physiol.* 1969; 217(1): pp. 65–72.

[23] De Geest H, Levy MN, Zieske H, Lipman RI. Depression of ventricular contractility by stimulation of the vagus nerves. *Circ Res.* 1965; 17: pp. 222–35.

[24] Randall WC, Wechsler JS, Pace JB, Szentivanyi A. Alterations in myocardial contractility during stimulation of the cardiac nerves. *Am J Physiol.* 1968; 214: 1205–12.

[25] Guyton AC, Jones CE, Coleman TG. *Circulatory Physiology: Cardiac Output and its Regulation.* Figure 18-1, p. 307. WB Saunders, London. 1973.

[26] Levy MN, Zieske H. Autonomic control of cardiac pacemaker activity and atrioventricular transmission. *J Appl Physiol.* 1969; 27: pp. 465–70.

[27] Case RB, Berglund E, Sarnoff SJ. Ventricular function: II. Quantitative relationship between coronary flow and ventricular function with observations on unilateral failure. *Circ Res.* 1954; 2: pp. 319–25.

[28] Case RB, Berglund E, Sarnoff SJ. Ventricular function. VII. Changes in coronary resistance and ventricular function resulting from acutely induced anemia and the effect thereon of coronary stenosis. *Am J Med.* 1955; 18(3): pp. 397–405.

[29] Rosenbueth A, Alanis J, Rubio R, Pilar G. Relations between coronary flow and work of the heart. *Am J Physiol.* 1961; 200: pp. 243–46.

[30] Stone HL, Bishop VS, Guyton AC. Cardiac function after embolization of coronaries with microspheres. *Am J Physiol.* 1963; 204: pp. 16–20.

[31] Guyton AC. Determination of cardiac output by equating venous return curves with cardiac response curves. *Physiol Rev.* 1955; 35(1): pp. 123–29.

[32] Guyton AC, Abernathy B, Langston JB, Kaufmann BN, Fairchild HM. Relative importance of venous and arterial resistances in controlling venous return and cardiac output. *Am J Physiol.* 1959; 196(5): pp. 1008–14.

[33] Guyton AC, Coleman TG. Long-term regulation of the circulation: interrelationships with body fluid volumes. In Reeves EB, Guyton AC (eds). *Physical Bases of Circulatory Transport: Regulation and Exchange.* WB Saunders, Philadelphia. 1967, pp. 179–201.

[34] Guyton AC, Jones CE, Coleman TG. *Circulatory Physiology: Cardiac Output and its Regulation.* Figure 17-1, p. 286. WB Saunders, London. 1973.

[35] Selkurt EE. The relation of renal blood flow to effective arterial pressure in the intact kidney of the dog. *Am J Physiol.* 1946; 147: pp. 537–49.

[36] Guyton AC, Jones CE, Coleman TG. *Circulatory Physiology: Cardiac Output and its Regulation.* Figure 17-2, p. 290. WB Saunders, London. 1973.

[37] Guyton AC, Lindsey AW, Abernathy B, Langston JB. Mechanism of the increased venous return and cardiac output caused by epinephrine. *Am J Physiol.* 1957; 192: pp. 126–130.

[38] Polosa C, Rossi G. Cardiac output and peripheral blood flow during occlusion of carotid arteries. *Am J Physiol.* 1961; 200: pp. 1185–90.

[39] Groom AC, Lofving BM, Rowlands S, Thomas HW. The effect of lowering the pulse pressure in the carotid arteries on the cardiac output in the cat. *Acta Physiol Scand.* 1962; 54: pp. 116–27.

[40] Corcondilas A, Donald DE, Shepherd JT. Assessment by two independent methods of the role of cardiac output in the pressor response to carotid occlusion. *J Physiol.* 1964; 170: pp. 250–62.

[41] Iriuchijima J, Soulsby ME, Wilson MF. Participation of cardiac sympathetics in carotid occlusion pressor reflex. *Am J Physiol.* 1968; 215(5): pp. 1111–14.

[42] Allison JL, Sagawa K, Kumada M. An open-loop analysis of the aortic arch barostatic reflex. *Am J Physiol.* 1969; 217: pp. 1576–84.

[43] Richardson TQ, Fermoso JF, Pugh GO. Effect of acutely elevated intracranial pressure on cardiac output and other circulatory factors. *J Surg Res.* 1965; 5: pp. 319–22.

[44] Sagawa K, Taylor AE, Guyton AC. Dynamic performance and stability of cerebral ischemic pressor response. *Am J Physiol.* 1961; 201: pp. 1164–72.

[45] Sagawa K, Carrier O, Guyton AC. Elicitation of theoretically predicted feedback oscillation in arterial pressure. *Am J Physiol.* 1962; 203: pp. 141–46.

[46] Guyton AC, Jones CE, Coleman TG. *Circulatory Physiology: Cardiac Output and its Regulation*. Figure 20-4, p. 362. WB Saunders, London. 1973.

[47] Brough RB Jr, Cowley AW Jr, Guyton AC. Quantitative analysis of the acute response to haemorrhage of the renin–angiotensin–vasoconstrictor feedback loop in areflexic dogs. *Cardiovasc Res*. 1975; 9: pp. 722–33.

[48] Tipayamontri U, Young DB, Nuwayhid BS, Scott RE. Analysis of the cardiovascular effects of arginine vasopressin in conscious dogs. *Hypertension*. 1987; 9: pp. 371–78.

[49] Crowell JW, Guyton AC. Evidence favoring a cardiac mechanism in irreversible hemorrhagic shock. *Am J Physiol*. 1961; 201: pp. 893–96.

[50] Crowell JW, Read WL. In vivo coagulation; a probable cause of irreversible shock. *Am J Physiol*. 1955; 183: pp. 565–69.

[51] Crowel JW, Sharpe GP, Lambright RL, Read WL. The mechanism of death after resuscitation following acute circulatory failure. Surgery. 1955; 38: pp. 696–702.

[52] Cowell JW, Smith EE. Oxygen deficit and irreversible hemorrhagic shock. *Am J Physiol*. 1964; 206: pp. 313–16.

[53] Jones CE, Crowell JW, Smith EE. A cause–effect relationship between oxygen deficit and irreversible hemorrhagic shock. *Surg Gynecol Obstet*. 1968; 127: pp. 93–96.

[54] Jones CE, Crowell JW, Smith EE. Significance of increased blood uric acid following extensive hemorrhage. *Am J Physiol*. 1968; 214: pp. 1374–77.

[55] Crowell JW, Guyton AC. Further evidence favoring a cardiac mechanism in irreversible hemorrhagic shock. *Am J Physiol*. 1962; 203: pp. 248–52.

[56] Guyton AC. Kidneys and fluids in pressure regulation. Small volume but large pressure changes. *Hypertension*. 1992; 19(1 Suppl): pp. 12–18.

[57] Granger JP. Pressure natriuresis. Role of renal interstitial hydrostatic pressure. *Hypertension*. 1992; 19(1 Suppl): pp. 19–17.

[58] Hall JE, Guyton AC, Brands MW. Control of sodium excretion and arterial pressure by intrarenal mechanisms and the renin–angiotensin system. In Laragh JH, Brenner BM (eds). *Hypertension: Pathophysiology, Diagnosis and Management*. Raven Press, New York. 1995.

[59] Sawka MN, Convertino VA, Eichner ER, Schnieder SM, Young AJ. Blood volume: importance and adaptations to exercise training, environmental stresses, and trauma/sickness. *Med Sci Sports Exerc*. 2000; 32(2): pp. 332–48.

[60] Nagashima K, Mack GW, Haskell A, Nishiyasu T, Nadel ER. Mechanism for the posture-specific plasma volume increase after a single intense exercise protocol. *J Appl Physiol*. 1999; 86: pp. 867–73.

[61] Yang RC, Mack GW, Wolfe RR, Nadel ER. Albumin synthesis after intense intermittent exercise in human subjects. *J Appl Physiol.* 1998; 84: pp. 584–92.

[62] Klausen K, Andersen LB, Pelle I. Adaptive changes in work capacity, skeletal muscle capillarization and enzyme levels during training and detraining. *Acta Physiol Scand.* 1981; 113(1): pp. 9–16.

[63] Andrew GM, Guzman CA, Becklake MR. Effect of athletic training on exercise cardiac output. *J Appl Physiol.* 1966; 21: pp. 603–8.

[64] Ekblom B, Astrand PO, Saltin B, Stenberg J, Wallstrom B. Effect of training on circulatory response to exercise. *J Appl Physiol.* 1968; 24: pp. 518–28.

[65] Hickson RC, Foster C, Pollock ML, Galassi TM, Rich S. Reduced training intensities and loss of aerobic power, endurance, and cardiac growth. *J Appl Physiol.* 1985; 58: pp. 492–99.

[illegible] Wan [illegible] [illegible] [illegible] substantia [illegible] subjects. *J* [illegible] pp. 54 [illegible]

[illegible] [illegible] [illegible] lateral [illegible] [illegible] ration and [illegible] [illegible] *Phys* [illegible] 2001, pp. 7 [illegible]

[illegible] Blake NR, Effects of [illegible] [illegible] [illegible] *Physiol* [illegible] [illegible] 9.

[illegible] [illegible] [illegible] [illegible] pp. [illegible]

[illegible] [illegible] [illegible] pp. 491–99.

Author Biography

David B. Young was born in Pittsburgh, Pennsylvania, in 1945. He graduated from Shady Side Academy in Pittsburgh in 1963, completed his undergraduate study at the University of Colorado in 1967, earned his Ph.D. degree in physiology from Indiana University in 1972, then began a two-year post-doctoral fellowship studying cardiovascular physiology with Arthur C. Guyton at the University of Mississippi Medical Center in Jackson, Mississippi. Dr. Young joined the faculty in the Department of Physiology and Biophysics at the University of Mississippi Medical Center in 1974 as an instructor and rose in rank to become a professor in 1980. He became professor emeritus in 2001. In 1970, he was awarded a National Institutes of Health (NIH) Individual Predoctoral Fellowship, an NIH Post-Doctoral Fellowship in 1972, and an NIH Research Career Development Award in 1977. Throughout his career, Dr. Young's research concerning the control and cardiovascular effects of potassium, vascular disease, and hypertension was supported by several grants from the NIH. He has published more than 100 articles in peer-reviewed journals, contributed chapters to several books, and authored a book concerning his research findings related to the cardiovascular protective effects of potassium. Dr. Young served in numerous review and consulting positions including being an associate editor (*American Journal of Physiology: Regulatory, Integrative and Comparative Section*) from 1995 to 2001, consultant (*Stedman's Medical Dictionary*, 27th and 28th editions), and chairperson (NIH HL/NAHAS Special Study Section), January and July 1986. During his career, he directed training of seven doctoral students and numerous post-doctoral fellows and he was awarded two patents for inventions related to artificial mechanical hearts. Currently, Dr. Young continues to publish and consult on topics related to his research interests in cardiovascular physiology and disease and the regulation of potassium and its cardiovascular effects. Dr. Young has been married since 1965 to his wife, Susan. They have three adult children and three grandchildren and divide their time between homes in Jackson, Mississippi, and near Basalt, Colorado. His interests include outdoor sports and activities, traveling, and photography.

Author Biography